GEOENGINEERING OUR CLIMATE?

If the detrimental impacts of human-induced climate change continue to mount, technologies for geoengineering our climate – i.e. deliberate modifying of the Earth's climate system at a large scale – are likely to receive ever greater attention from countries and societies worldwide. Geoengineering technologies could have profound ramifications for our societies, and yet agreeing on an international governance framework in which even serious research into these planetary-altering technologies can take place presents an immense international political challenge.

In this important book, a diverse collection of internationally respected scientists, philosophers, legal scholars, policymakers, and civil society representatives examine and reflect upon the global geoengineering debate they have helped shape. Opening with essays examining the historic origins of contemporary geoengineering ideas, the book goes on to explore varying perspectives from across the first decade of this global discourse since 2006. These essays methodically cover the practical and ethical dilemmas geoengineering poses; the evolving geoengineering research agenda; the challenges geoengineering technologies present to current international legal and political frameworks; and differing perceptions of geoengineering from around the world. The book concludes with a series of forward-looking essays, some drawing lessons from precedents for governing other global issues, others proposing how geoengineering technologies might be governed if/as they begin to emerge from the lab into the real world.

This book is an indispensable resource for scientists, activists, policymakers, and political figures aiming to engage in the emerging debate about geoengineering our climate.

Jason J. Blackstock is the founding head of the Department of Science, Technology, Engineering, and Public Policy at University College London.

Sean Low is a research associate at the Institute for Advanced Sustainability Studies in Potsdam, Germany.

SCIENCE IN SOCIETY SERIES

Series Editor: Steve Rayner

Institute for Science, Innovation and Society, University of Oxford

Editorial board: *Jason J. Blackstock, Bjorn Ola Linner, Susan Owens, Timothy O'Riordan, Arthur Petersen, Nick Pidgeon, Dan Sarewitz, Andy Stirling, Chris Tyler, Andrew Webster, Steve Yearley*

The Earthscan Science in Society Series aims to publish new high-quality research, teaching, practical and policy-related books on topics that address the complex and vitally important interface between science and society.

Assessing the Societal Implications of Emerging Technologies
Anticipatory Governance in Practice
Evan Michelson

Aid, Technology and Development
The Lessons from Nepal
Edited by Dipak Gyawali, Michael Thompson and Marco Verweij

Climate Adaptation Policy and Evidence
Understanding the Tensions between Politics and Expertise in Public Policy
Peter Tangney

Cities and the Knowledge Economy
Promise, Politics and Possibilities
Tim May and Beth Perry

Institutional Capacity for Climate Change Response
A New Approach to Climate Politics
Theresa Scavenius and Steve Rayner

Geoengineering Our Climate?
Ethics, Politics, and Governance
Edited by Jason J. Blackstock and Sean Low

GEOENGINEERING OUR CLIMATE?

Ethics, Politics, and Governance

Edited by Jason J. Blackstock and Sean Low

Routledge
Taylor & Francis Group

LONDON AND NEW YORK

First published 2019
by Routledge
2 Park Square, Milton Park, Abingdon, Oxon OX14 4RN

and by Routledge
711 Third Avenue, New York, NY 10017

Routledge is an imprint of the Taylor & Francis Group, an informa business

British Library Cataloguing-in-Publication Data
A catalogue record for this book is available from the British Library

Library of Congress Cataloging-in-Publication Data
Names: Blackstock, Jason J., editor. | Low, Sean, editor.
Title: Geoengineering our climate?: ethics, politics and governance / edited by Jason J Blackstock and Sean Low.
Description: Abingdon, Oxon; New York, NY: Routledge, 2019. | Series: Science in society series | Includes bibliographical references and index.
Identifiers: LCCN 2018021128 | ISBN 9781849713733 (hbk) | ISBN 9781849713740 (pbk) | ISBN 9780203485262 (ebk)
Subjects: LCSH: Environmental geotechnology. | Environmental geotechnology—Moral and ethical aspects. | Climate change mitigation. | Climate change mitigation—Moral and ethical aspects.
Classification: LCC TD171.9 .G454 2019 | DDC 628—dc23
LC record available at https://lccn.loc.gov/2018021128

ISBN: 978-1-84971-373-3 (hbk)
ISBN: 978-1-84971-374-0 (pbk)
ISBN: 978-0-203-48526-2 (ebk)

Typeset in Bembo
by Deanta Global Publishing Services, Chennai, India

CONTENTS

FIGURES

TABLES

BOXES

ACRONYMS AND ABBREVIATIONS

AOSIS	Alliance of Small Island States
AR4	Fourth Assessment Report (IPCC)
AR5	Fifth Assessment Report (IPCC)
BBC	British Broadcasting Corporation
BECCS	bioenergy production with carbon dioxide capture and storage (CDR)
BPC	Bipartisan Policy Center
CBD	Convention on Biological Diversity
CCS	carbon dioxide capture and storage
CDR	carbon dioxide removal
CERN	European Organization for Nuclear Research
CGIAR	Consultative Group on International Agricultural Research
CIAG	Climate Institute Asilomar Group
CLRTAP	Convention on Long Range Transboundary Air Pollution
CMIP	Coupled Model Intercomparison Project
CNS-ASU	Center for Nanotechnology in Society at Arizona State University
COP	Conference of the Parties (e.g. of UNFCCC, or other international convention)
DAC	direct air capture (CDR)
Earth Summit	UN Convention on Environment and Development
EDF	Environmental Defense Fund
ELSI	ethical, legal, and social implications
ENGO	environmental non-governmental organisation
ENMOD	UN Convention on the Prohibition of Military or Any Other Hostile Use of Environmental Modification Techniques
EOR	enhanced oil recovery
E-PEACE	Eastern Pacific Emitted Aerosol Cloud Experiment
FAR	First Assessment Report (IPCC)
FAR	Fraction Attributable Risk

GeoMIP	Geoengineering Model Intercomparison Project
GGF	Global Governance Futures programme
GHG	greenhouse gas(es)
GMO	genetically modified organisms
HGP	Human Genome Project
HoC S&T	UK House of Commons Science and Technology Committee
HSRC	Haida Salmon Restoration Project
HUGO	Human Genome Organisation
IAGP	Integrated Assessment of Geoengineering Proposals
ICSU	International Council for Science
IGY	International Geophysical Year
IMF	International Monetary Fund
IMO	International Maritime Organization
IMPLICC	Implications and Risks of Engineering Solar Radiation to Limit Climate Change
IOPC	International Oil Pollution Compensation
IP	intellectual property
IPCC	Intergovernmental Panel on Climate Change
ITER	International Thermonuclear Experimental Reactor
LOHAFEX	an OIF experiment conducted by Indian and German research institutes
London Convention	Convention on the Prevention of Marine Pollution by Dumping of Wastes and Other Matter
London Protocol	Protocol to the Convention on the Prevention of Marine Pollution by Dumping of Wastes and Other Matter
MDG	Millennium Development Goals
NERC	Natural Environment Research Council
NGO	non-governmental organisation
NNI	US National Nanotechnology Initiative
NOAA	National Atmospheric and Oceanic Administration (US)
OIF	ocean iron fertilisation
Ops	Oxford Principles
OTA	Office of Technology Assessment, now closed (US)
ppm	parts per million
Rio Declaration	Rio Declaration on Environment and Development.
SAR	Second Assessment Report (IPCC)
SIDS	small island developing states
SPI	sulphate/stratospheric particle injection (SRM)
SPICE	Stratospheric Particle Injection for Climate Engineering
SPM	Summary for Policy Makers (IPCC Assessment Reports)
SRM	solar radiation management or sunlight reflection methods
SRMGI	Solar Radiation Management Governance Initiative
TAR	Third Assessment Report (IPCC)
TWAS	the academy of sciences for the developing world

UNCLOS	UN Convention on the Law of the Sea
UNEP	UN Environmental Programme
UNFCCC	United Nations Framework Convention on Climate Change
USGAO	Government Accountability Office (US)
VEC	Virgin Earth Challenge (CDR)
WCRP	World Climate Research Program
WG I,II,III	Working Group(s) I, II, III (IPCC)
WIPO	World Intellectual Property Organization
WMO	World Meteorological Organization
WTO	World Trade Organization

CONTRIBUTORS

Cheryl Lea Anderson, PhD, director of LeA International Consultants, also affiliated with the University of Hawai'i Social Science Research Institute, conducts research and planning projects on climate and disaster risk management in the Asia-Pacific region, with attention to gender, indigenous knowledge, and socioeconomic aspects of risk reduction, adaptation, and resilience.

Mulugeta Mengist Ayalew serves as a director of climate change affairs in the Office of the Prime Minister, Ethiopia. He is also a research assistant professor of law in Mekelle University.

Bidisha Banerjee is a social ecologist and leadership coach. She is completing *Superhuman River: A Biography of Ganga* due from Aleph Books in New Delhi in 2018. She has been involved in geoengineering governance issues since the 2010 Asilomar conference.

Rob Bellamy is a James Martin Research Fellow in the Institute for Science, Innovation and Society at the University of Oxford. His research focuses on the interactions between global environmental change and society, particularly in relation to decision-making, public participation, innovation governance, and risk perception.

Jason J. Blackstock is the co-founder (2013) and former head (2013–18) of the Department of Science, Technology, Engineering and Public Policy at University College London where he remains a faculty member. With a unique background spanning physics, technology development, and international affairs, Jason is a leading international scholar, educator, and policy adviser on the interface between science and global public policy.

Chad Briggs is a principal consultant with GlobalINT, senior fellow at the Institute for Environmental Security in Brussels and adjunct professor at Johns Hopkins University. He specialises in translation of complex scientific data into risk assessments and strategic planning.

Holly Jean Buck is a post-doctoral researcher at the Institute of the Environment and Sustainability, University of California Los Angeles, where her research interests combine geography and narratives in social science analysis, and include agroecology, energy landscapes, land use change, algal biofuels and synthetic biology, new media, and science and technology studies.

Rose Cairns is a research fellow at the Science Policy Research Unit at the University of Sussex and coordinates the UK Economic and Social Research Council's Nexus Network. She engages in interdisciplinary research for sustainability, science and technology policy, knowledge politics, and environmental governance.

Wylie Carr is a social scientist with the US Fish and Wildlife Service in Atlanta, Georgia. As a PhD candidate at the University of Montana he explored how populations vulnerable to impacts from climate change view geoengineering. The research presented in this book is based on his research as a graduate student.

Ying Chen is a senior research fellow at the Institute for Urban and Environmental Studies, Chinese Academy of Social Sciences, deputy director of CASS Research Center for Sustainable Development and professor at CASS Graduate School. She has been engaged in environmental economics and sustainable development, as well as energy and climate change policy.

Paul J. Crutzen is an emeritus scholar at the Max Planck Institute for Chemistry in Mainz, Germany, where he had previously served as scientific director (1983-1985) and as scholar (1980-2000). Crutzen won the Nobel Prize for Chemistry in 1995 (shared with M. J. Molina and F. S. Rowland) for his work demonstrating the effects of nitrous oxide on ozone layer depletion.

Jack Doughty was formerly a researcher in climate engineering governance in the Department of Science, Technology, Engineering and Public Policy (STEaPP) at University College London.

ETC Group (the Action Group on Erosion, Technology and Concentration) works to address the socioeconomic and ecological issues surrounding new technologies that could have an impact on the world's poorest and most vulnerable people. It investigates ecological erosion, the development of new technologies, and global governance issues including corporate concentration and trade in technologies.

James Rodgers Fleming is the Charles A. Dana Professor of Science, Technology, and Society at Colby College and a research associate at the Smithsonian National Air and Space Museum. He has written extensively on the history of weather, climate, technology, and the environment including social, cultural, and intellectual aspects.

Rider W. Foley is an assistant professor in the Department of Engineering and Society at the University of Virginia.

Johannes Gabriel is the founder and director of Foresight Intelligence, a systemic consultancy for strategic foresight and organisational learning. He is also adjunct professor at Johns Hopkins University's SAIS, teaching scenario planning in its Global Risk program. He a non-resident fellow with the Global Public Policy Institute in Berlin and a team member of its Global Governance Futures programme.

Florent Gasc is geo information officer at the United Nations Logistical Base and a monitoring and evaluation consultant for projects and programmes dealing with climate change, environment, and development. He was formerly at the African Climate Policy Centre.

Arunabha Ghosh is a public policy professional, adviser, author, columnist, and institution builder. He is the founder-CEO of the Council on Energy, Environment and Water (CEEW), among the 'world's leading climate think tanks. He is the co-author/editor of four books, has work experience in 42 countries, and previously worked at Princeton, Oxford, UNDP, and WTO.

David H. Guston is founding professor and director of the School for the Future of Innovation in Society at Arizona State University, where as director of the NSF-funded Center for Nanotechnology (2005–2011) he led the development of real-time technology assessment and the anticipatory governance of emerging technologies.

Steven P. Hamburg is chief scientist at the Environmental Defense Fund, where he ensures the scientific integrity of EDF's positions and programmes, facilitates research collaborations, and helps identify emerging science relevant to EDF's mission. He is an ecosystem ecologist specialising in biogeochemistry and climate impacts on forest structure and function.

Alex Hanafi is senior manager of multilateral climate strategy and a senior attorney in EDF's International Climate Program, where he coordinates research and advocacy programmes designed to promote policies and build institutions that effectively reduce greenhouse gas emissions around the globe.

Joshua B. Horton is research director, geoengineering, with the Keith Group at the Harvard Kennedy School and is affiliated with Harvard's Solar Geoengineering

Research Program. He conducts research on geoengineering policy and governance issues, including institutions and regime design, decision-making, ethics, and geopolitics.

David Keith is Gordon McKay Professor of Applied Physics in the School of Engineering and Applied Sciences and professor of public policy at the Harvard Kennedy School, and is faculty director for Harvard's Solar Geoengineering Research Program. Keith is founder of Carbon Engineering, a start-up company developing industrial scale technologies for capture of CO_2 from ambient air to produce transportation fuels from solar power.

Ben Kravitz is a climate scientist in the Atmospheric Sciences and Global Change Division at the US Department of Energy"s Pacific Northwest National Laboratory. His research primarily focuses on using climate models to answer fundamental physical science questions regarding the effects of solar geoengineering.

Robert E. Kopp is director of the Institute of Earth, Ocean, and Atmospheric Sciences and a professor in the Department of Earth and Planetary Sciences at Rutgers University-New Brunswick. His research focuses on understanding uncertainty in past and future climate change, with major emphases on sea-level change and on interactions between physical climate change and the economy.

Tim Kruger is James Martin Research Fellow and programme manager of the Oxford Geoengineering Programme at the Oxford Martin School. He leads a group exploring proposed geoengineering techniques and the governance mechanisms required to ensure that any research in this field is undertaken in a responsible way.

Mark G. Lawrence is scientific director at the Institute for Advanced Sustainability Studies in Potsdam, Germany. He conducts and leads research focusing on environmental sustainability issues such as the impacts and mitigation of short-lived climate-forcing pollutants, and on the potential impacts, uncertainties, and risks of climate engineering.

Penehuro Fatu Lefale, director of LeA International Consultants, works as a climate and policy advisor for Pacific island countries. Lefale has worked for the Meteorological Service of New Zealand, the World Meteorological Organization, the Secretariat of the Pacific Environment Programme, and the National Institute of Water and Atmospheric Research of New Zealand.

Timothy M. Lenton is professor of climate change and Earth system science at the University of Exeter, where he researches the coupled evolution of life and planet Earth, develops and uses Earth system models, and studies the prospects for early warning of climate tipping points.

Sean Low is a research associate at the Institute for Advanced Sustainability Studies in Potsdam, Germany. His research explores how scientific activity uses conceptions of the future to shape current actions in research and policy, focusing on modelling, foresight, gaming, and analogies to other emerging technologies.

Duncan McLaren is a research fellow at Lancaster University, working on the risk of mitigation deterrence from carbon geoengineering. Amongst other roles, he is also an advisor to the Virgin Earth Challenge, and the Leverhulme Centre for Climate Change Mitigation. Previously, he was chief executive of Friends of the Earth Scotland from 2003 to 2011.

Janot Mendler de Suarez is a visiting research fellow with Boston University's Pardee Center for the Study of the Longer-Range Future, technical advisor and Caribbean focal point with the Red Cross Red Crescent Climate Centre, currently consulting with the World Bank Caribbean Disaster Risk Finance Technical Assistance programme.

Juan B. Moreno-Cruz is an assistant professor in the School of Economics at the Georgia Institute for Technology. He focuses on the interaction of energy systems, technological change, and climate policy, and is developing a new set of theoretical and empirical tools to study energy-system transitions in order to inform energy and environmental policy.

David R. Morrow is a faculty fellow with the Forum for Climate Engineering Assessment at American University and a visiting fellow at the Institute for Philosophy and Public Policy at George Mason University, studying philosophical and policy issues surrounding climate justice and the governance of climate engineering.

Oliver Morton is senior editor for essays and briefings at *The Economist*. He is the author of three books: *Mapping Mars: Science, Imagination and the Birth of a World* (2002); *Eating the Sun: How Plants Power the Planet* (2007), and *The Planet Remade: How Geoengineering Could Change the World* (2015).

Simon Nicholson is assistant professor in the School of International Service and director of the Global Environmental Politics Program at American University. His work focuses on global food politics and the politics of emerging technologies. He is co-founder of the Forum for Climate Engineering Assessment, a scholarly initiative of the School of International Service.

Michael Oppenheimer is the Albert G. Milbank Professor of Geosciences and International Affairs in the Woodrow Wilson School and the Department of Geosciences at Princeton University. His interests include science and policy of the atmosphere, particularly climate change and its impacts on the ice sheets and sea level, on the risk from coastal storms, and on patterns of human migration.

Andy Parker is project director of the Solar Radiation Management Governance Initiative, an international, NGO-driven project that seeks to expand the global conversation around the governance of SRM geoengineering research. He has worked on geoengineering for more than nine years, including at the Institute for Advanced Sustainability Studies, the Harvard Kennedy School, and the Royal Society.

Arthur Petersen is professor and director of doctoral studies at the Department of Science, Technology, Engineering and Public Policy at University College London. Previously, he was a scientific adviser on environment and infrastructure policy within the Dutch government, and served as chief scientist of the PBL Netherlands Environmental Assessment Agency from 2011–2014.

Jesse L. Reynolds is Emmett/Frankel Fellow in Environmental Law and Policy at the University of California, Los Angeles School of Law and associated researcher with the Utrecht Centre for Water, Oceans, and Sustainability Law and Harvard's Solar Geoengineering Research Program. His monograph *The Governance of Solar Geoengineering* is forthcoming on Cambridge University Press.

Katharine L. Ricke is an assistant professor at the University of California – San Diego, with appointments in the School of Global Policy and Strategy and at Scripps Institution of Oceanography. Her research combines quantitative modelling and large data set analysis techniques applied to physical and social systems.

Alan Robock is a distinguished professor of climatology in the Department of Environmental Sciences at Rutgers University, where his research focuses on geoengineering, climatic effects of nuclear weapons, soil moisture variations, the effects of volcanic eruptions on climate, and the impacts of climate change on human activities. Robock also serves as editor of reviews of *Geophysics*.

Daniel Sarewitz is professor of science and society at the School for the Future of Innovation in Society and co-director and co-founder of the Consortium for Science, Policy, and Outcomes (CSPO) at Arizona State University. His work focuses on revealing and improving the connections between science policy decisions, scientific research, and social outcomes.

Dane Scott is the director of the Mansfield Ethics and Public Affairs Program at the University of Montana and associate professor of ethics in the College of Forestry and Conservation. Scott's primary research interests are in the ethics and philosophy of technology, most recently on geoengineering and biotechnology.

Pablo Suarez is associate director for research and innovation at the Red Cross Red Crescent Climate Centre, where he leads initiatives linking applied knowledge with humanitarian work, and explores new threats and opportunities on climate risk management (such as climate engineering, financial instruments, or participatory games). He has consulted for some 20 humanitarian and development organisations, working in more than 50 countries.

Bronislaw Szerszynski is reader in the Department of Sociology, Lancaster University, UK. His research draws on the social sciences, humanities, arts, and natural sciences in order to place contemporary changes in the relationship between humans, the environment, and technology in the longer perspective of human and planetary history.

Michael Thompson is senior outreach manager for the Carnegie Climate Geoengineering Governance Initiative (C2G2). Before joining C2G2, Thompson was a co-founder of the Forum for Climate Engineering Assessment, and researcher at the Smithsonian Institution in Washington, DC and the Energy and Resources Institute in Delhi, India.

Jeff Tollefson is a US correspondent for *Nature* in New York, where he covers energy, climate, and the environment. He has been the recipient of fellowships from the Alicia Patterson Foundation and the Knight Science Journalism Program at MIT.

Nancy Tuana is the founding director of Penn State's Rock Ethics Institute and DuPont/Class of 1949 Professor of Philosophy and Women's Studies. Tuana is a philosopher of science who specialises in issues of ethics and science. She was a CoPI of the NSF funded Network on Sustainable Risk Management (SCRiM) and the NSF funded interdisciplinary project Visualizing Forest Futures.

Gernot Wagner is a research associate at Harvard's School of Engineering and Applied Sciences, lecturer on environmental science and public policy, co-director of Harvard's Solar Geoengineering Research Program, and co-author of *Climate Shock* (Princeton University Press, 2015).

Weili Weng received her PhD in economics from the Institute for Urban and Environmental Studies, Chinese Academy of Social Sciences. Her research interests are in climate change policies, decision-making under risk and uncertainty, and international governance of geoengineering.

ACKNOWLEDGEMENTS

With over 30 contributions and 50 authors, the creation of this book – conceived as an accessible entry point to the human dimensions of climate geoengineering – was a collective endeavour spanning more than half a decade. The contents are indicative of the vibrancy and variety of early perspectives surrounding geoengineering, and the authorship is representative of the diverse communities and geographies debating the development of geoengineering technologies with passion and rigour. We are grateful to all our contributing authors – and to all those who engaged in discussions along the way – for enriching this increasingly important global dialogue.

The journey to this book began with the participants of the *Geoengineering Our Climate? Ethics, Politics and Governance* conference held in January 2012 in Ottawa (Canada), which was kindly supported by the Centre for International Governance Innovation. It continued in the years since through an online working paper series of the same name (www.geoengineeringourclimate.com) that provided a forum for authors to share and get feedback on the perspectives they were developing. This uncommon arrangement of publishing early drafts of contributions to this book for comment and debate was generously supported by our publisher (Earthscan for Routledge), and enabled by a partnership of editorial institutions spanning four continents:

- The Institute for Science, Society and Innovation (InSIS) at Oxford University
- The Department of Science, Technology, Engineering and Public Policy (STEaPP) at University College London
- The Consortium on Science, Policy and Outcomes (CSPO) at Arizona State University
- The Institute for Advanced Sustainability Studies (IASS) in Germany
- The Council on Energy, Environment and Water (CEEW) in India

- The Research Centre for Sustainable Development of the Chinese Academy of Social Sciences, and
- The Brazilian Research Group on International Relations and Climate Change.

In addition to improving and enriching the contents of this book, the working paper process also signalled a shared ethos of the contributing community – specifically, our collective belief that climate geoengineering technologies are of such potentially transformative importance to the Earth and to humanity, that diverse ideas and perspectives about these technologies need to be disseminated and debated as widely as possible at every stage in their development. We are grateful to these institutions for lending their support, their visibility, and their credibility to this project – and especially to our editors at Earthscan who took the risk on this experimental approach.

Leading members of these institutions – Steve Rayner, Clark Miller, Jack Stilgoe, Mark Lawrence, Arunabha Ghosh, Pan Jiahua, and Eduardo Viola – all contributed at various points as associate editors of the working paper series, enhancing the contributing authors' work and shaping the flow of this book. Jeff Tollefson, Jeff Goodell, and Oliver Morton contributed to the process by interviewing authors, creating accessible introductory videos for a number of papers, and expanding the reach and impact of this work. Numerous colleagues – academic, policy, and civic – provided comments on early drafts of working papers, improving the rigour and richness of the final book. We are deeply grateful to everyone who contributed their effort and expertise throughout this project.

Finally, we are grateful to our home institutions – University College London and the Institute for Advanced Sustainability Studies – for enabling us to devote time to this project over the last half decade. Without their support, this book would not have been possible.

1

GEOENGINEERING OUR CLIMATE

An emerging discourse

Jason J. Blackstock and Sean Low

In 2006 Nobel laureate Paul Crutzen, renowned for his contributions to understanding and addressing our planetary ozone layer crisis, published a paper that drew new attention to old ideas (Crutzen, 2006). Within it he suggested that, if it remained unchecked, the mounting impacts of human-induced climate change could become so damaging that humanity might need to explore ways of intentionally cooling the Earth's climate to counteract it. This paper, coming as it did amid mounting international concern about anthropogenic climate change, catalysed what has since evolved into an expanding global discussion about *geoengineering* – the deliberate, large-scale alteration of the Earth's climate system.

The specific idea that Crutzen described for intentionally altering our climate system – that of injecting sulphate aerosols into the stratosphere to act as tiny reflectors, reducing the amount of sunlight reaching the ground and thereby cooling the planet – had been around for decades (Keith, 2000). In recent years, the proposed engineering of planetary sunshades has been discussed alongside a second kind of intervention: the creation of new carbon sinks, either through novel technological means or by enhancing the capacity of natural systems to do so. Respectively termed *solar radiation management* (SRM) and *carbon dioxide removal* (CDR) approaches, these two distinct categories are still often referred to together as geoengineering. However, this is not without protest from those who argue that the different objectives, characteristics, and trajectories in research and policy of various SRM and CDR approaches outweigh the value of discussing them together as deliberate and sustained attempts at climate management. This is an unfinished and evolving debate, which we return to briefly later in this introduction.

For now, we note that such ideas for intentionally engineering the Earth's climate to suit human needs and aspirations have a long and complex intellectual history – a history that the authors of the four chapters presented in Part I of this book explore using a range of lenses and examples. Prior to Crutzen's provocative article, however, geoengineering concepts were widely dismissed as hubristic and far-fetched, belonging more to the realm of futuristic science fiction than modern science. Rigorous scientific research into geoengineering concepts was scarce and limited almost entirely to theoretic analyses and calculations. More importantly, serious public or political conversations about actually developing and potentially deploying geoengineering technologies were non-existent.

By the mid-noughties, a clear scientific consensus had emerged on climate change: humans were the primary cause; the risks to humanity were numerous, serious, and escalating; and near-term actions to mitigate the human causes of climate change were essential. Unfortunately, the politics of the mid-noughties did not align with the scientific consensus. Fierce political debates about climate policies raged, spurred by concerns about possible economic costs and social disruptions from changing the industries and lifestyles underlying the greenhouse gas emissions causing climate change. These debates generated widespread concern among scientific and environmental communities that the political will needed to effectively mitigate climate change might not emerge in time to avoid serious, potentially catastrophic damage to future populations around the world. Published within this context, Crutzen's reframing of geoengineering – as perhaps being able to forestall or ameliorate the future ravages of climate change, should humanity not act quickly enough to avoid them – attracted unprecedented new attention to these concepts.

This reframing was not uncontroversial – and it remains heatedly contested today. The notion of intentionally altering the Earth's climate to counteract the climatic change that humanity was already unintentionally causing soon became the subject of considerable contemplation. In Part II of this book, authors of the seven chapters present and explore both their own and others' framings of climate engineering ideas from a diversity of perspectives. How should such planet-altering technologies be thought about and discussed? What are the ethics surrounding their development and potential use? Where should they fit among the suite of other potential responses to the risks of climate change? Are they viable even as a last resort? Widely varying implicit and explicit answers to these and many similar questions are evident throughout the contributions to this book.

Alongside the burgeoning ethical and philosophical discussions, new scientific research agendas also emerged. As reflected on by Crutzen himself in this book, following his 2006 article "the openness of the [scientific] community towards research" into potential geoengineering concepts "changed rapidly" (Lawrence and Crutzen, Chapter 13). The ensuing decade would see numerous scientific assessments of geoengineering technologies by internationally respected scientific organisations and groups (e.g. Novim, 2009; Royal Society, 2009; APS, 2011; IPCC, 2011; National Research Council, 2015a, 2015b; EuTRACE, 2015), and the emergence of numerous modelling studies, technology design and development activities, and even some early field experiments. The last decade of research, along with perspectives on what the next decade might hold, are explored by the authors of the five chapters presented in Part III of this book.

Despite this increased attention, progress towards the technological development of deployable geoengineering technologies has remained limited over the last decade. No country has yet taken the bold step of formally authorising or significantly funding the technological development of geoengineering alternatives. While the scientific research has repeatedly reinforced the potential for geoengineering technologies to reduce many impacts of rising temperatures, certain scenarios could result in different regional outcomes (see Irvine et al., 2016 for an accessible summary of the Earth systems science and modelling of SRM). Similar to climate change, the planetary-scale nature of prospective geoengineering impacts raises challenging political and legal questions of responsibility and sovereignty for the international community to grapple with. Whether or not existing global institutions and still emerging frameworks can effectively manage these issues is the subject of both expanding scholarship and debate in some international forums. In the four chapters presented in Part IV of this book, authors explore how climate geoengineering has been, or may soon be, tackled by existing international institutions.

Of course, the lack of serious political action should not be interpreted as a lack of serious political perspectives. If developed or even just formally explored, geoengineering technologies could be a game-changer in international climate politics, and possibly even in global politics generally. Unsurprisingly, diverse perspectives have quickly emerged on how geoengineering alternatives could or should evolve, shaped at least in part by national and regional political contexts. The seven chapters in Part V of this book attempt to provide a snapshot of the range of political perspectives that have been articulated over the last decade, with authors from diverse backgrounds and perspectives making compelling arguments about how geoengineering could be framed and managed. No one book could hope to do justice to the full diversity of viewpoints that exist on geoengineering. What these contributions – and others in this book – do capture is how rapidly the notion of geoengineering has infused various and sundry policy discussions and debates worldwide.

Will expanding scientific research into geoengineering lead to technological development and deployment of geoengineering technologies? Should it? How will the global discussion about geoengineering evolve? Whose perspectives will be taken into account as it does? How should humanity ultimately govern geoengineering? There are no simple answers to any of these questions, although proposals for how to approach them abound. The final section of this book (Part VI) is devoted to authors exploring what the effective governance of still emerging geoengineering research and technologies might look like, though these chapters are far from the final word. Like the other contributions in this volume, these chapters advance and further clarify the governance challenges facing humanity with the exploration of an idea as profound as potentially *geoengineering our climate*.

Coming over a decade after Paul Crutzen's original paper, our hope is that the contributions throughout this book will help readers formulate their own perspectives on this complex and evolving debate, and about which ethical, legal, and political questions are most important for them to further explore and engage. We hope this helps empower our readers to decide for themselves how they can most effectively contribute to the growing global discourse about the role geoengineering technologies should or should not play within humanity's response to climate change.

This book aims to provide readers with an accessible and multidisciplinary introduction to the complex human dimensions of geoengineering. Along the way, it also attempts to situate and contextualise the emergence of geoengineering as a global discourse. As hopefully conveyed by the narrative above, our aim is to retrace a number of important ideas as they developed, focusing on broad sweeps – historical antecedents, framing concepts, trends in early research and institutional governance, and geographic or sector-based perspectives on the enterprise of managing the climate – rather than on the technical details of individual approaches.

We do so by capturing a particular window in the evolution of the geoengineering discourse – specifically, the period following Crutzen's catalytic 2006 article, during which the concept of geoengineering was beginning to emerge into wider scientific and political conversations around the world. The acknowledgements on page xxii detail how the journey to this book began in 2012, with authors contributing early drafts to an online working paper series that ran till 2015, stimulating broad discussions and debates (*geoengineeringourclimate.com*). Over the course of the last several years, many ideas shared in those draft papers and captured in this book have gone on to be further developed and written up, influencing and shaping the global geoengineering discourse in myriad ways.

In assembling the final contents of this book, we note that most of the concerns, challenges, and questions discussed herein remain just as relevant today as they did when first drafted. This is a testament both to the quality of contributing authors and to the timeless nature of some of the deep human questions evoked by the idea of geoengineering. At the same time, we must qualify that it is not our intent to provide a current affairs summary of all the ideas within an ever-evolving debate. Any attempt to do so would invariably be out of date in numerous ways by the time the book was printed. No anthology can capture the full spectrum of thought that has emerged in the past decade, and for every idea that remains resilient, another has been contested or even overturned.

Nevertheless, we attempt to address these issues through a number of measures. To help readers navigate the diverse, sometimes divergent, perspectives in this book on what geoengineering is, what it means, and what it could be and might do, a brief introduction is provided for each section. These introductions provide a light-touch narrative weaving the contributions together. They also attempt to locate the chapters in subsequent developments of the issues that they address, by pointing out topics or evolutions in the discourse that are not covered by the book's contributions or by the editors' organisation thereof, accompanied by more recent papers and resources. Our aim with this approach is to continue the ethos that underpinned the original working paper series – i.e. to support a vibrant and ongoing discussion about emerging geoengineering technologies, rather than attempting to create a static repository.

Two final notes are necessary; the first on organisation. The reader will notice that chapters are of two different lengths, and reflect differing emphases on academic scholarship, practitioner experience, and personal or institutional perspective. This mirrors the original organisation of the online series into *working papers* and *opinion articles*. Working papers were longer and comparatively neutral overviews (~6000 words) of ethical, political, and governance questions, which have become Chapters 2, 12, 20, 31, and 34 within this volume. Conversely, opinion articles were shorter pieces (~2000 words) that provided flexibility to cover subjective commentaries, experimental ideas, or short focused case studies. Given the rich variety of contributions this approach elicited, we have not followed a traditional format of using longer chapters to precede and frame shorter ones in each section. Rather, we have interspersed both kinds of contributions amongst each other to facilitate a narrative flow that is partly topical and partly chronological – from the historical and conceptual at the front end, to intervening politics, and to proposed governance frameworks for the future towards the end. We also group all works cited in a collective bibliography at the end of the book, instead of by chapters. This is because contributions cite many common texts (the Royal Society's 2009 report is a prime example); it also reduces clutter in a book with 40 contributions. In the end, our editorial structuring of this volume represents only one narrative framing, and there is no correct way to read or frame the contributions in this book. We encourage readers to see this book as a collection of diverse voices, each bringing to light different aspects of the global geoengineering discourse.

Our second note is on terminology. For readers less familiar with this discourse, it is important to note that *geoengineering* is a broad and heavily contested umbrella term, encompassing a diverse range of technical proposals for intentionally altering the Earth's climate system. As previously noted, two broad categories of geoengineering concepts are discussed: (i) ways to remove carbon dioxide molecules from the atmosphere, generally referred to as *carbon dioxide removal* (CDR) technologies; and (ii) ways to reflect sunlight away from the Earth to cool the climate, generally referred to as *solar radiation management* (SRM) technologies. For readers interested in the scientific details of geoengineering proposals, we encourage you to look at the accessible scientific summaries and reviews that have been published over the last decade (see "Key assessment reports" below, particularly Royal Society, 2009; EuTRACE, 2015; and National Research Council, 2015a and 2015b).

There have been many attempts to systematise or alter the nomenclature surrounding geoengineering concepts. Key assessments since the Royal Society's seminal 2009 report have (re)titled geoengineering as climate engineering (EuTRACE, 2015), climate remediation (BPC, 2012), or climate intervention

(National Research Council, 2015a and 2015b), the lattermost also having taken the novel step of producing separate reports for SRM and CDR. Some call for eliminating the umbrella term completely and re-situating individual technological proposals, or at least as the broad suites of sunshades or carbon sinks, within new schemes for categorising all climate strategies (e.g. Heyward, 2013; Boucher et al., 2014; Pereira, 2016). Much specialised debate continues behind the scenes on how to categorise new and old approaches under the umbrella term or its constituent suites. Can afforestation, reforestation, biochar, and other land-use measures that preceded geoengineering be reframed as such? Why can marine cloud brightening be considered a sunlight reflection method, while cirrus cloud thinning cannot?

At various points throughout the development of this project we debated these and other ways to address the definitional challenge on approaches that are simultaneously overlapping, distinct, and evolving. We settled on the pragmatic choice of allowing the contributing authors to use their own judgement and provide their own definitions, with the (by now standard) caveat that they note explicitly whether their arguments refer specifically to either SRM or CDR, to individual approaches within those suites, or to the broader enterprise of deliberate climate geoengineering. While at first this may seem a difficult choice for the reader, it reflects the reality that the evolving geoengineering discourse has been (and continues to be) beset with definitional challenges and uncertainties. Each advocate, critic, or commentator builds their arguments with the definition(s) they find most suitable, leaving listeners or readers to make their own interpolation across to other contexts or discussions. As a result, the definitional variety within this book adds a dimension that we believe ultimately enriches the reader's understanding of the nuances and complexities of this discourse.

As geoengineering has edged over the last decade ever closer to the public and political mainstream, the state of our knowledge about both the science and the human dimensions of the potential technologies has continued to evolve. No doubt it will continue to do so. The contributions in this book are indicative of the variety and vibrancy, the rigour and passion that have characterised the emerging conversations and debates about geoengineering technologies. We hope you find them as rich and rewarding as we have!

KEY ASSESSMENT REPORTS ON GEOENGINEERING

American Physical Society (2011): *Direct Air Capture of CO2 with Chemicals. A Technology Assessment Report for the American Physical Society Panel on Public Affairs*

→ https://www.aps.org/policy/reports/assessments/upload/dac2011.pdf

Asilomar International Conference on Climate Intervention Technologies (2010): *The Asilomar Conference Recommendations on Principles for Research into Climate Engineering Techniques*

Bipartisan Policy Center, US (2012): *Geoengineering: A National Strategic Plan for Research on the Potential Effectiveness, Feasibility, and Consequences of Climate Remediation Technologies*

→ http://bipartisanpolicy.org/library/report/task-force-climate-remediation-research

Bracmort, K. and Lattanzio, R.K., Congressional Research Service, US (2013): *Geoengineering: Governance and Technology Policy*

→ http://www.fas.org/sgp/crs/misc/R41371.pdf

EuTRACE (2015): *The European Transdisciplinary Assessment of Climate Engineering: Removing Greenhouse Gases from the Atmosphere and Reflecting Sunlight away from Earth*

→ http://www.eutrace.org/

House of Commons, Science and Technology Committee, UK (2010): *The Regulation of Geoengineering*

→ http://www.publications.parliament.uk/pa/cm200910/cmselect/cmsctech/221/22102.htm

Intergovernmental Panel on Climate Change (2011): *Summary of the Synthesis Session*. In: *IPCC Expert Meeting Report on Geoengineering*

→ https://www.ipcc-wg1.unibe.ch/publications/supportingmaterial/EM_GeoE_Meeting_Report_final.pdf

National Research Council, US (2015a): *Climate Intervention: Carbon Dioxide Removal and Reliable Sequestration*

→ https://doi.org/10.17226/18805

National Research Council, US (2015b): *Climate Intervention: Reflecting Sunlight to Cool Earth*

→ https://doi.org/10.17226/18988

Novim (2009): *Climate Engineering Responses to Climate Emergencies*

→ http://arxiv.org/pdf/0907.5140

Rickels, W., et al., commissioned by the Federal Ministry of Education and Research, Germany (2011): *Large-Scale Intentional Interventions into the Climate System? Assessing the Climate Engineering Debate*

→ http://www.kiel-earth-institute.de/scoping-report-climate-engineering.html

(*continued*)

Royal Society, UK (2009): *Geoengineering the Climate: Science, Governance and Uncertainty*

→ http://royalsociety.org/policy/publications/2009/geoengineering-climate/

US Government Accountability Office (2010a): *Climate Change: Preliminary Observations on Geoengineering Science, Federal Efforts, and Governance Issues*

→ https://www.gao.gov/assets/130/124271.pdf

US Government Accountability Office (2010b): *Climate Change: A Coordinated Strategy Could Focus Federal Geoengineering Research and Inform Governance Efforts*

→ http://www.gao.gov/products/GAO-10-903

US Government Accountability Office (2011): *Climate Engineering: Technical Status, Future Directions, and Potential Responses*

→ http://www.gao.gov/products/GAO-11-71

Working Paper Series and Other Resources

Climate Engineering News (News-aggregation website for the latest publications, initiatives, and meetings)

→ http://www.climate-engineering.eu/news.html

Climate Geoengineering Governance (CGG) Working Paper Series

→ http://geoengineering-governance-research.org/cgg-working-papers.php

Geoengineering Our Climate? Ethics, Politics and Governance Working Paper Series

→ https://geoengineeringourclimate.com/

Governance Scoping Initiatives (operational in 2018)

Carnegie Climate Governance Geoengineering Initiative (C2G2)

→ https://www.c2g2.net/

Forum for Climate Engineering Assessment (FCEA)

→ http://ceassessment.org/

Solar Radiation Management Governance Initiative (SRMGI)

→ http://www.srmgi.org/

Research Programs (operational in 2018)

Carbon Dioxide Removal Intercomparison Project (CDR-MIP)

→ https://www.kiel-earth-institute.de/CDR_Model_Intercomparison_Project.html

Climate Engineering in Science, Society and Politics, Institute for Advanced Sustainability Studies, Germany

→ https://www.iass-potsdam.de/en/research/climate-engineering-science-society-and-politics

Climate Engineering Research Programme, Linköping University, Sweden

→ https://liu.se/en/research/linkoping-university-climate-engineering-research-programme-luce

Exploring the Potentials and Side Effects of Climate Engineering, Norway (EXPECT)

→ https://expected.bitbucket.io/

German Research Foundation (DFG) Priority Programme (SPP) 1689, Germany

→ http://www.spp-climate-engineering.de/focus-program.html

Geoengineering Model Intercomparison Project (GeoMIP)

→ http://climate.envsci.rutgers.edu/GeoMIP/

Geoengineering Research Governance Project (GRGP)

→ http://www.ucalgary.ca/grgproject/

Greenhouse Gas Removal from the Atmosphere Research Programme, UK

→ http://www.nerc.ac.uk/research/funded/programmes/ggr/

Mechanisms and Impacts of Geoengineering, China

→ http://www.china-geoengineering.org/

Solar Geoengineering Research Program, Harvard University, US

→ https://geoengineering.environment.harvard.edu/

Research Programs (completed)

Climate Geoengineering Governance (CGG)

→ http://geoengineering-governance-research.org/

(*continued*)

European Transdisciplinary Assessment of Climate Engineering (EuTRACE)

→ http://www.eutrace.org/

Implications and Risks of Engineering Solar Radiation to Limit Climate Change (IMPLICC)

→ https://implicc.zmaw.de/index.php?id=551

Integrated Assessment of Geoengineering Proposals (IAGP)

→ http://www.iagp.ac.uk/node.html

Stratospheric Particle Injection for Climate Engineering (SPICE)

→ http://www.spice.ac.uk/

PART I

Historical context

Introduction

The breadth and depth of human musings on our potential for mastery over nature is far beyond the ability of any one book to capture. Nevertheless, it is important to recognise that the imagining of geoengineering concepts is only one more step in a long journey of humans altering their environmental surroundings to suit their purposes. From the advent of agriculture to the ongoing global urbanisation of the 21st century, humanity has been constantly transforming the planetary environment we inhabit. At each step, scientists and scholars have both conceived and developed new tools of mastery, and reflected on the philosophical and moral quandaries these tools have given rise to.

The contributions in this first section of the book continue in this tradition. The aim of this collection of chapters is to situate and contextualise the contemporary exploration and debates about climate geoengineering concepts in the long arc of philosophic thought surrounding humanity's relationship with nature. Dane Scott begins this process in Chapter 2, examining how conceptions of geoengineering situate within the longstanding philosophic traditions of technological optimism, pessimism, and pragmatism. This is followed closely by Bronislaw Szerszynski, who in Chapter 3 uses four compelling caricatures of historic figures to explore how geoengineering concepts intersect with and challenge fundamental precepts of religion and the sacred role of nature.

A vast literature on the philosophical and ethical underpinnings of geoengineering has been usefully compiled at the University of Montana's Ethics of Geoengineering Online Resource Center (see Bibliography), exploring its dimensions in areas ranging from the strategic (arguments for or against research or deployment; influences upon mitigation and adaptation efforts) to the practical (conducting stakeholder engagement or establishing governance mechanisms);

from the conceptual (theology and religion, gender, or anthropocentricism) to the temporal (intergenerational considerations and engaging with the future). We also note that investigation of the philosophical and ethical aspects of geo-engineering has been argued to have undergone two waves. The first treated geoengineering as a broad concept of planetary intervention, while calling for interdisciplinary deliberations and stakeholder engagement. The second, still taking shape at the time of this book's publication, calls for a more fine-grained analysis of ethical dimensions unique to particular SRM and CDR approaches, and for integrating these discussions with those surrounding mitigation, adaptation, and wider climate strategies (Baatz et al., 2016). Scott's and Szerszynski's chapters can be seen as part of this first wave, whereas Tuana's contribution later in this book (Part II, Chapter 12), focused on sulphate particle injection, aligns more with the second.

From these more philosophic foundations, the other two chapters in this section turn to historical examples of technologies arguably related to contemporary geoengineering concepts. Such comparative analogies are rife in the geoengineering literature, though deployed with varying degrees of rigour – ranging from broad comparisons used for polemic purposes, to concerted proofs-of-concept that frame the challenges and governance needs for various geoengineering approaches by emphasising their similarities to past or concurrent debates in global governance. The comparisons provided by the authors in these chapters follow the latter course, articulating comparative analyses that promote deep consideration by the reader.

In Chapter 4, James Fleming provides a short history of technological attempts and proposals over the last two centuries to modify the weather. From mid-19th century attempts to control rain, to mid-20th century ideas for intentionally altering the Earth's climate using nuclear bombs, Fleming's chapter reminds us that current proposals for geoengineering in response to climate change have complex, chequered origins, and warns us that the militaristic, securitised logic of Cold War era weather modification remains a spectre to be concerned by. In Chapter 5, Oliver Morton places contemporary climate geoengineering proposals in the broader context of humanity's impact on other aspects of the Earth's systems, suggesting that humanity's engineering of the nitrogen cycle in the 20th century might provide a compelling "existence proof" comparison for how humanity may adopt a similar approach to the Earth's climate in the 21st century.

A full mapping of similar robust comparative analogies deployed in geoengineering literature and debates does not currently exist. Nonetheless, we would commend interested readers to at least two other prominent comparisons found in the literature. First is the comparison to the governance of other emerging technology areas, such as nanotechnology and synthetic biology. These comparisons have particularly been used in recent years to suggest that, similar to other areas of emerging technology where public engagement has proven critical, discussions of geoengineering technologies might better reflect public concerns if they were democratised beyond expert and technocratic circles (e.g. Chapter 34 in this volume by Foley, Guston, and Sarewitz; Stilgoe, 2015). The second prominent comparison is to a variety of eco-modernist or eco-pragmatist technologies, such as genetically modified organisms (GMOs). These comparisons have particularly been used to argue that human civilisation's footprint on the planet should not be rolled back as much as shaped via innovative interventions that are ultimately beneficial across our entwined natural and societal systems (e.g. Brand, 2009).

2

PHILOSOPHY OF TECHNOLOGY AND GEOENGINEERING

Dane Scott

William James begins his classic philosophical work, *Pragmatism*, with remarks by G.K. Chesterton. Chesterton says that the most important thing we can know about a person is their philosophy, their "theory of the cosmos". He asserts that it is more important for a landlord to know a renter's philosophy than income; that in war, while it is crucial to know the enemy's numbers, it is even more important to know the enemy's philosophy. James affirms Chesterton's remarks; he comments that each of us has a philosophy and that "the most important and interesting thing about [us] is the way in which it determines the perspective in [our] several worlds" (James, 1981, p. 7). The astonishing idea of geoengineering, that some group(s) of humans could attempt to intentionally manage the Earth's climate through large-scale technological means, has philosophical significance. As a leading scientific expert on geoengineering, David Keith (2009a), writes, "Deliberate planetary engineering would be a new chapter in humanity's relationship with the earth". How people interpret and react to geoengineering will largely depend on what James broadly terms their "philosophy".

In its fifth report, the Intergovernmental Panel on Climate Change included a significantly expanded assessment of geoengineering (Petersen, 2018). This has raised awareness of geoengineering and further moved the debate towards centre stage. Starkly conflicting philosophies of technology will play a role in shaping people's reactions to geoengineering in conflicting ways. How this debate plays out will be a major factor in determining the future of geoengineering research, governance, and possible deployment. This chapter, then, takes seriously the idea that it is important to know the philosophies that shape people's interpretations of, and reactions to, geoengineering. More specifically, this chapter will worry about what philosophies will ultimately help shape our collective response to geoengineering. And, it will inquire into what philosophy, or philosophies, would best serve to undergird our collective response to geoengineering.

In an effort to gain a full view on these concerns, I will serve as a guide of sorts in a brief tour of philosophy of technology – to get the lay of the land, if you will. The landscape of philosophy of technology has been mapped in several ways, and it is far more detailed and complicated than this brief tour might indicate. This tour will look at three contrasting philosophies of technology, which I will term: (1) technological progressivism, (2) technological pessimism, and (3) technological "pragmatism". From each of the three perspectives, important questions are asked about geoengineering.

To briefly clarify important terms: for the purposes of what follows, Chesterton's phrase "theory of the cosmos" is taken to mean something like what cultural theorists now call a "worldview". While he doesn't make this distinction, James's more narrow term "philosophy" will refer to more specific philosophical ideas, principles, and interpretative schemas that play a key role in shaping a person's worldview. The three philosophies of technology examined below play important roles in shaping worldviews. Individuals and groups with varying degrees of self-awareness, coherency, and self-criticism and who take positions on controversial technologies adopt at least elements of these three philosophies. Once again, one's philosophy of technology is an important factor in determining how one interprets and reacts to controversial technologies, like nuclear power, genetic engineering, and geoengineering.

Technological progressivism

Technological progressivism has the longest history of the settled territories in philosophy of technology. It is often traced to a history of philosophy that held that social progress was the inevitable outcome of scientific and technological development. For many people in Western societies, particularly in the United States, technological progressivism remains influential, even if its utopian dreams are now seen as an illusion.

From this perspective, the consistent application of science and technology is humanity's greatest hope for improving human life (Hanks, 2007, p. 2). The philosopher Hans Achterhuis (2001, p. 68) describes its essential assumptions as "purely instrumental, utterly neutral with respect to political and social choices. This social-political neutrality is said to result from the rational and universal character of technology". The most important feature to keep in mind from this view is the universal and instrumental character of technology as a tool for making life better and more efficient.

While most academic philosophers of technology have long abandoned this territory, technological progressivism remains influential in social debates over technologies. Perhaps one of the reasons this philosophy of technology remains influential is due to tremendous institutional inertia. Many of Western society's most powerful and wealthy institutions, research universities, national research centres, and technology companies, particularly in the United States, were created in the 20th century when this philosophy was in its ascendancy. These institutions have a vested interest in perpetuating this philosophy of technology, in whole or in part.

One of the most important thinkers who helped shape technological progressivism in the 20th century was Alvin Weinberg, director of the Oak Ridge National Laboratory from 1955 to 1973. A memorial essay for Weinberg states that, "more than any other scientist of his generation he communicated the meaning and intent of Big Science" (ORNL Review, 2002). Weinberg's importance is found in his advocacy for the philosophy of technological progressivism through "Big Science" and "technological fixes" – Weinberg coined both terms. During the second half of the 20th century, Weinberg played the role as public philosopher of technology, writing and speaking on the relationship between technology and society (Weinberg, 1967, p. 2).

Weinberg's belief in the power of Big Science to solve social problems leads to a pattern of thinking, captured by his influential and now infamous notion of the technological fix. Weinberg argued that while technological fixes cannot replace "social engineering" (i.e., politics and political philosophy), they should be used as a "positive social action" (Teich, 1993). Weinberg characterises a technological fix as the solution to a social or political problem that results from reframing it as a technological one. The major benefit of doing this is it reduces the seemingly insurmountable complexity of social and political problems to "beautiful and crisp technological solutions" (ibid.). Weinberg lists the major benefits of technological fixes. First, as noted, technological problems are much simpler than social problems; it is easier to define and identify solutions to technological problems. Second, technological problems do not have to deal with the complexity and unpredictability of the social world. Third, they provide policymakers with more options and additional means for addressing social problems. Finally, they can buy time until the problem can be dealt with on a deeper level.

Weinberg anticipates many of the contemporary criticisms of technological fixes. For example, in reducing the complexity of a problem by reframing it as an engineering puzzle, many important factors that can generate unintended consequences can be excluded. In "fixing" one problem, technological fixes often generate others. Also, as is commonly noted, technological fixes do not get at the "root causes" of a problem. However, from the view of Big Science, side effects involve risks and trade-offs, which can often be addressed with additional technologies. Further, it is often simply unrealistic to hope that some social problems can be dealt with through political means; technological fixes may be the best we can do at the time.

Another influential advocate of technological progressivism was Weinberg's friend and contemporary, the eminent physicist Freeman Dyson. In his book, *The Sun, the Genome and the Internet*, it is clear that Dyson shares Weinberg's paradigm of solving social problems through Big Science and technological fixes. Dyson (1999, p. 49) writes that his purpose for considering the relationship between technology and society is to look for "ways in which technology may contribute to social justice, to alleviate the differences between rich and poor, to the preservation of the earth". While acutely aware that modern technologies have a mixed record, sometimes generating unwanted social and environmental side effects, Dyson judges the overall trajectory of modern technologies to be progressive (ibid., p. 56). He is convinced that "new technologies offer the opportunity for making the world a happier place" (ibid., p. 61).

Geoengineering proposals are textbook examples of the Big Science-technological fix pattern of inquiry into social problems. Most discussions of geoengineering begin by stating that the reason scientists and policymakers are considering geoengineering research is the incapacity of political processes to deal with the climate problem. The overwhelming complexity within social and political systems has made global climate change an intractable problem. Yet when the climate crisis is viewed as an engineering puzzle, numerous potential technological solutions emerge – solutions that will restore the Earth's energy balance to pre-industrial levels. In the late 1970s, Dyson published an article (1977) introducing an early geoengineering proposal. The merits of Dyson's particular proposal are of little interest to the present discussion. What is important is to observe how a person's philosophy of technology provides a perspective on geoengineering. Given his philosophy of technology, it is not surprising the Dyson readily considers a technological fix to acute, dangerous climate change.

Dyson asks a speculative question that is being taken much more seriously 30+ years later. He writes: "Suppose that with the rising level of CO_2 we run into an acute ecological disaster. Would it then be possible for us to halt or reverse the rise in CO_2 within a few years by means less drastic than the shutdown of industrial civilization" (ibid.). Dyson speculates that it "should be possible in the case of a world-wide emergency to plant enough trees and other fast-growing plants to absorb the excess CO_2 and bring the annual increase to a halt" (ibid.). The purpose of his research was to provide rough calculations that demonstrate the economic and technical feasibility of such a plan. Dyson notes that this "climate engineering" plan would be a short-term emergency response that would buy time – a stated benefit of a technological fix – for the long-term solution of shifting away from a fossil fuel economy. The above discussion is not concerned with the merits of Dyson's proposal;[1] the point is to illustrate how one's philosophy of technology provides a perspective on geoengineering. Further, it could be argued that many of the people proposing geoengineering schemes share this basic view of technology: that it is progressive, universal, and instrumental. For some, this philosophy of technology leads them to look to geoengineering for technological fixes to provide at least a short-term solution for an entrenched, seemly insoluble social problem. Again, central to this position is the assumption that technologies are morally and politically neutral, that human intentions, good or bad, direct technologies.

This view of technology does provide tools to criticise technological developments. It offers a specific set of important critical questions to ask about controversial emerging technologies. One can question the motives behind the technology and ask about potential side effects. For example, most of the criticisms of geoengineering proposals are lists of potentially impure motives and unwanted side effects.[2] However, critical questions are restrained by this instrumentalist's view of philosophy of technology. From this view, the technology itself is morally and politically neutral. The kinds of criticisms of geoengineering from the next region of philosophy of technology are much deeper and more sweeping.

Technological pessimism

Deep philosophical critiques of technological progressivism populate this territory of philosophy of technology. These critics paint a dark picture of the increasing role of technology in contemporary culture. From their perspectives, human history is now on a dangerous trajectory where technology is destined to become the "determining and controlling influence on society and culture" (Verbeek, 2005, p. 11). One of the deepest critiques comes from German philosopher Martin Heidegger who "understands technology as a particular manner of approaching reality, a dominating and controlling one in which reality can only appear as raw material to be manipulated" (ibid., p. 10). Rather than being a tool guided by human intentions for good or ill, technology has become a malignant dominating force of culture and nature.

Cultural historian Leo Marx provides a summary of this philosophy of technology. He notes that it is popular among modern intellectuals to dismiss technological fixes to social and environmental problems. However, their criticisms treat the habitual impulse as an isolatable error, a mistaken way of thinking that can be easily corrected (Marx, 1983, p. 7). Marx warns that the problem runs too deep to be easily corrected. He writes: "Unfortunately, the dangerous idea of a technical fix is embedded deeply in what was, and probably is, our culture's dominant conception of history" (ibid.). The dangerous idea of a technological fix is part of a worldview that structures reality and how we interact with it. Marx calls the philosophical assumption of technological optimism a "logical abyss in our thinking", and responds to it by asserting that, "few arguments could be more useful today than one aimed at persuading the world that science and technology, essential as they are, cannot save us" (ibid.).

Another important source of this philosophy is environmental historian Lynn White's famous essay, "The Historical Roots of our Ecological Crisis" (1967). White's thesis is that the 20th century's environmental crisis is the result of the enormous power created by the union of science and technology along with a worldview that justifies the use of that power to dominate and control nature. White aims to undermine the view that technological power is essentially a benign and progressive force and the idea that humans have the right to dominate the Earth to satisfy our needs.

Looking at geoengineering in this light, Dale Jamieson (2010, p. 441) argues that climate change, let alone geoengineering, "violates a duty of respect for nature because it is a central expression of the human domination of nature".

From this view, "geoengineering demonstrates a culpable attitude of domination and is quite probably a 'paradigm of disrespect'" (Preston, 2011, p. 462). Geoengineering is more than a bad idea; it is an expression of a misguided philosophy of technology and a dysfunctional worldview. From this view, the fact that geoengineering proposals are starting to rise into consideration must seem tragically predictable. Jamieson gives expression to this perspective in a survey of possible responses to geoengineering. He writes that, "even if [geoengineering] were successful, it would still have the bad effect of reinforcing human arrogance and the view that the proper human relationship to nature is one of domination" (Jamieson, 1996).

Philosopher Clive Hamilton has written several essays that are critical of geoengineering. These essays contain many of the themes of technological pessimism. In a 2011 essay, Hamilton takes direct aim at the instrumentalism of technological progressivism as part of the worldview that has led to geoengineering. He writes:

> The objectification of the earth means regarding it as a collection of resources that have instrumental value only, that is, values only as means to human ends. Viewing the earth in instrumental terms, so that the ethics of acting on it are to be judged purely by their effects, requires a certain wonderlessness and estrangement from the earth.
>
> *(Hamilton, 2011, p. 3)*

Hamilton provides several interesting arguments from Earth science to discourage "thinking of the world as a systematic totality that we can know and control" (ibid.). Echoing Heidegger, he urges us to abandon the technological thinking that sees the "world as comprised of resources at our disposal that can be grasped with our minds and manipulated with technology" (ibid.). Rather than research geoengineering, Hamilton, voicing the positive vision of technological pessimism, urges humanity to adopt a humility that will allow us to live with nature.

Technological pessimism is a largely critical philosophy. It is highly influential among many environmental thinkers and, in general, is more prominent in Europe than in North America. Philosophers and humanists in this territory of philosophy of technology see their task as shaking people from the entrenched view on reality as shaped by technological progressivism. Their goal is to point out the dangerous blindness and distortions created by instrumentalism and progressivism that lead to the domination of nature through scientific technology. If we are to be saved we must envision a world where humans live humbly as a part of nature. This will call for new political, economic, and technological systems.

This dark perspective of technological society does provide important insights. The most significant is the challenge to take a deeply self-critical perspective by questioning the intellectual foundations that have led to a seeming addiction to technological fixes. However, the wholesale rejection of technological fixes, such as geoengineering in an acute climate crisis, may cause great human suffering and loss of biodiversity. Moreover, when this perspective becomes the exclusive view on these issues, it can create an ideological fanaticism.

Technological "pragmatism"

The final stop on this quick tour of philosophy of technology and geoengineering is technological "pragmatism".[3] This territory has been carved out over the last three decades by thinkers who saw limitations and distortions in the earlier views. The thinkers in this area are diverse and are yet to have the widespread cultural impact of the two previous views. In this region, philosophers provide more specific analyses of a range of particular technologies using a variety of philosophical approaches (Ihde, 2003).

There are several philosophers in this area whose works could be used to shed light on geoengineering and philosophy of technology. However, for the sake of brevity, I will discuss just one issue in light of the ideas of one philosopher in this final section. The philosopher is Albert Borgmann and the issue is the common criticism that technological fixes, like geoengineering, do not target an issue's underlying cause. The purpose of this brief discussion is to provide an illustration of a new perspective for asking important questions about geoengineering that go beyond the familiar, if important, questions that can be asked from the perspectives of technological progressivism and pessimism.

One standard criticism of technological fixes like geoengineering is they do not get at the "root" of the problem. Most problems have multiple causes and the cause people identify as the root cause to address is generally a choice based on various criteria.[4] It is hard to say that there is an ultimate cause to any problem. However, the root cause criticism is not making a point about causal theory and strategic action. It is not saying that technological fixes are ineffective; they often are as good as can be done given the circumstances. Rather, the moral intuition behind this criticism is that in using technological fixes we are avoiding important engagements in the world that ought to be required to solve a problem. Technological fixes can unburden us of efforts and interactions in the world that are morally significant, whether by developing virtues or promoting justice.

One of Borgmann's important insights is captured in his notion of the device paradigm. An essential feature of modern technologies is they tend to make things available for us without imposing burdens. Borgmann (1984) writes that, "something is available in this sense if it has been rendered instantaneous, ubiquitous, safe and easy". An important consequence of making things readily available is it lessens our engagements with the natural world and others. Borgmann uses the example of central heating to describe the trade-offs that come with the device paradigm. To be warned, the point of the analysis is not to advocate going back to an earlier time when homes were heated with wood but to introduce a pattern of analysis that allows one to become aware of important engagements that are lost when new technological devices are introduced. With the advent of central heating, many laborious, time consuming, and sometimes dangerous tasks involved in heating a home with wood were taken over by devices (ibid., p. 42). It is easy to imagine the numerous interactions with the natural world and others required by gathering wood and tending the fire. Moreover, the hearth created a focal point that gathered the family together for warmth through a common activity. This is in sharp contrast to the absence of engagement required by the technological system – from the thermostat in the home to the power plant in some unknown location – that keeps modern homes at a constant, comfortable temperature.

Again, Borgmann is not counselling that we give up the convenience of central heating. Technologies can allow for civilisations to advance by reducing our daily burdens and free us for other worthwhile activities. Rather, the loss of the warmth of the hearth as a focal point for family life demonstrates the kinds of losses that can be experienced because of the device paradigm. Devices bring many benefits, but we must be circumspect about how they lessen our engagement with others and the world. For Borgmann the task of "philosophers of technology is to point out what happens when technology moves beyond lifting genuine burdens and starts freeing us of burdens that we should not want to be rid of" (Wood, 2003). This perspective allows us to ask a new set of questions about specific technologies. In looking at various geoengineering proposals, then, one of the many questions to ask is: is this particular technology lifting us of burdens that we should not be rid of?

To illustrate this point, it will be helpful to compare two geoengineering proposals that are in sharp relief. It is important to keep in mind that these example technologies are chosen solely on the merit of providing ready contrast for the purposes of quickly illustrating this type of analysis. The first example is marine cloud brightening. An artist's depiction commonly accompanies articles that illustrate this proposal. In that illustration there is a fleet of automated, sleek, futuristic ships with large vertical cannons. The cannons are spewing atomised seawater into marine clouds to increase the number of water droplets and their albedo or reflective capacity. The point of brightening the clouds is to offset some of the increased solar radiation being trapped by greenhouse gases. The contrasting example is the brightening of dark urban surfaces, particularly roofs and roads. One article on this proposal is accompanied by a photograph of two workers standing on a flat roof in a dense urban landscape rolling white paint onto the black roof. The point of painting the roof white is the same as brightening the marine clouds. Of course, in order to be geoengineering, both proposals would have to be intentionally aimed at changing the Earth's climate and at a scale large enough to effect measurable change. To be warned, these two proposals may or may not be quantitatively comparable: that is a scientific question. However, these quantitative issues do not affect the point being made in terms of illustrating how the device paradigm might be used to discuss the moral and political significance of various geoengineering proposals. So, a more qualified question is: all things considered, does this particular proposal take away burdens we ought not to be rid of?

The failure of politics is most often cited for justifying geoengineering research. And, to a significant degree, the lack of engagement by citizens in democratic societies contributes to the lack of political success. The inability of politics to deal with an issue is a classic justification for a technological fix. This justification may be convincing; dangerous climate change seems imminent. At present, politics do not seem up to the task of acting in time to avoid very serious consequences for the Earth's poor and the tremendous loss of the Earth's biodiversity. However, no geoengineering proposal can ultimately save us from this political failure: at best geoengineering proposals are technological fixes capable of buying time until effective action can be taken. Moving economies away from fossil fuels while allowing developing economies to grow will involve hard social and political work. If we were to research geoengineering, it would be wise to put our effort towards technologies that do not worsen the problem of disengagement. To repeat the question: all things being equal, which of these two technologies, marine cloud brightening or the brightening of urban surfaces, would increase people's engagement with the issue?

On the one hand, the unmanned ships that would implement marine cloud brightening would do their work far from people's consciousness. The fleets of cloud brightening machines would be invisible to all but a few technicians. While implementation would be controversial and generate discussion for a time, once implemented it would require no engagement by a great majority of people. Moreover, marine cloud brightening would be far from politically neutral. Arguably, it would lend itself to centralised, technocratic governance. It is likely that initiating this technology would create a political fight that could have the positive result of engaging more people in the climate issues. However, once implemented, for all intents and purposes, the public could ignore it unless things went wrong. On the other hand, the brightening of urban surfaces could lend itself to more consistent immediate engagement by people. The activity of lightening roofs and roads would require a high degree of public engagement at the personal and local levels. On a personal level, one could imagine some people willingly buying more reflective roofing material while others complain bitterly. Either way, people would be directly engaged in the issue. On a civic level, one could imagine a contentious city council meeting as citizens debated brightening public surfaces. These changes would be no doubt highly contentious, generating numerous discussions between citizens on the climate issue. Moreover, brightened surfaces would serve as a more immediate reminder of the climate issue than marine cloud brightening could. Many questions could be raised from this terse illustration. However, it is clear that marine cloud brightening would more closely follow the device paradigm than would the brightening of urban surfaces.

In sum, when viewed from the device paradigm, research into proposals that strongly reinforce this paradigm should be avoided, as they would likely make it harder to create the necessary moral resolve that will ultimately be required to power the political will to address the climate problem. Those that do more to increase engagement and understanding of the issue would be better choices for research. Furthermore, proposals that naturally favour technocratic, centralised governance should be avoided, while those that favour more democratic, decentralised governance would be better choices for investigation. One could easily imagine taking the simple illustration above as a model for discussing other geoengineering proposals. In particular, it would be interesting to contrast various proposals that remove carbon from the atmosphere with those that attempt to block or reflect solar radiation.

Conclusion

The purpose of this brief tour of philosophy of technology is to demonstrate, in a very broad way, how these three philosophies of technology provide perspectives on geoengineering. The instrumentalism of technological progressivism would likely cause people to look past the moral and political significance of geoengineering proposals. From this view, the discussion would focus on side effects, efficiency, and efficacy. On the other hand, the determinism of technological pessimism would reject geoengineering proposals as a colossal instance of the domination of nature. From this view, geoengineering only furthers a dysfunctional worldview that needs to be abandoned. At the beginning of this chapter, I asked: what philosophy or philosophies of technology will ultimately help shape our collective response to geoengineering? Unfortunately, it is highly likely that these two highly limited philosophies of technology will dominate the geoengineering debate. I also asked: what philosophy or philosophies should shape our collective response to geoengineering? It would be wise to focus on more recent trends in the philosophy of technology. Borgmann's philosophy provides just one example of how this can be done. What I am suggesting is that an important component of the debate over geoengineering research should be over specific proposals and their moral and political significance. The view of geoengineering from technological "pragmatism" adds a fuller perspective by going beyond discussions of efficiency, effectiveness, potential side effects, or sweeping condemnation. Unfortunately, it has yet to have the culture and worldview shaping effects of the previous views. Further research on philosophy of technology and geoengineering is needed to have a well-informed debate over geoengineering.

First published online as a working paper in 2013.

Notes

1 In his contrarian essay on climate change in 2008, Dyson updates this technological fix to dangerous climate change. He predicts that in just a few short decades biotechnologists will be able to create "genetically engineered carbon-eating trees" (Dyson, 2008). Dyson speculates that these "carbon-eating trees could convert most of the carbon that they absorb from the atmosphere into some chemically stable form and bury

it underground. Or they could convert the carbon into liquid fuels and other useful chemicals" (ibid.). This geoengineering scheme would require replanting one quarter of the world's forests in carbon-eating varieties of the same species that exist naturally in these locations. This would be a win–win technological solution as the genetically engineered "forests would be preserved as ecological resources and as habitats for wildlife, and the carbon dioxide in the atmosphere would be reduced by half in about fifty years" (ibid.).

2 For example, while some geoengineering schemes might address the problem of rising temperatures, they would allow the associated problem of ocean acidification due to rising concentrations of CO_2 in the atmosphere to increase in severity. In addition, there are some worries that stratospheric aerosols would have the unintended consequence of contributing to the depletion of the ozone layer. Also, geoengineering projects will not act uniformly on the climate; they will likely change precipitation patterns, which in turn impact agricultural production. This could benefit some countries while harming others (Robock, 2008).

3 In the scholarly literature this new area has been labelled the "empiricist turn" (Brey, 2010). I am using the term "pragmatism" to mean being more realistic. The idea is that the earlier philosophies of technology, while containing many important insights, were either too optimistic about technological progress or too pessimistic about its negative social consequences. The general idea is that the more time thinkers have spent considering the social, ethical, and political implications of technology, the more realistic those considerations have become.

4 The cause of illness we choose to address is often an implicit or explicit social choice. For example, much illness can be attended to at the level of public health, preventative measures, or at the level of biomedicine, pharmaceuticals.

3

GEOENGINEERING AND THE SACRED

A brief history in four characters

Bronislaw Szerszynski

Geoengineering, the large-scale technological control of climate processes in order to offset anthropogenic global warming, is at this stage merely a prospective technology, yet it is already a controversial one. The idea of attempting to bring about deliberate large-scale changes in a complex, dynamic, chaotic system such as the atmosphere is highly ambitious. Like rDNA biotechnology and other new and emerging technologies before it, it is likely to provoke objections from opponents which are religious or implicitly religious in nature; for example, that it is "playing God" – overstepping the proper and safe limits of human freedom and control.

But it would be a mistake to see the influence of religious or cultural beliefs merely on the side of those who are sceptical or critical of new technologies. Cultural and religious ideas can also be seen as shaping the very idea of geoengineering. In *Nature, Technology and the Sacred*, I argued that in order to understand our current "technological condition" we have to locate it within the context of long historical trends and transformations in Western religion (Szerszynski, 2005). Here, I want to show how that kind of analysis can be applied to geoengineering, to reveal how it is an aspiration that has roots that are a lot longer than the modern period (Fleming, 2010), and is entangled with enduring themes about human agency and nature in Western cultural history.

In order to do this I will take a selection of "characters" from history who were ascribed the power to control atmospheric phenomena, and explore them as models for the "maker of climates" that is imagined in geoengineering discourse.

My first character is the *sacred king*. In many pre-modern societies the role of the king was not just political but also ritual: kings were seen as maintaining a vital link between human societies and the cosmos. In Mesopotamia, for example, the strict observance by the king of the annual cycle of festivals was seen as crucial for the maintenance of life and prosperity (Frankfort et al., 1949). The sacred kings of Tara in Ireland were married to the Earth goddess, which was believed to help ensure a constant supply of agricultural products. If the climate changed, threatening famine, this was a sign that the king had failed to maintain the goodwill of the goddess, which would often result in him suffering ritual mutilation, sacrifice, and burial (Dalton, 1970; Kelly, 2012).

The second character is the *cunning woman* or man of medieval European culture, who performed various divinatory, medical, and religious roles within peasant communities. According to the popular culture of the time, a range of rites, premonitions, and spells could be deployed in order to understand and manipulate the invisible bonds between things in the natural and supernatural worlds, including influencing the weather (Muchembled, 1985, pp. 71–79). This often led cunning folk to attract accusations of witchcraft and satanism, of using weather magic to create storms to sink ships or unseasonal hail and frost to blight crops (Briggs, 2002, pp. 78–79). Yet beliefs in weather magic in Europe predated such accusations of diabolism, and were manifest in pre-Christian pagan practices of appealing to spirits and deities to spare a village from storm or to bring rain to the crops (Horsley, 1979, pp. 86–87).

My third character is *the magus* of Renaissance Europe. Influenced by the texts of the occult European Hermetic tradition, these elite thinkers believed that they could use their knowledge of the natural and occult virtues of objects to attain great powers over natural phenomena. As Heinrich Cornelius Agrippa put it, magicians could "attain power over nature, and perform operations so marvellous, so sudden, so difficult, by reason of which ... the stars are disturbed, deities are compelled, the elements are made to serve" (Easlea, 1980, p. 99).

Finally, let us consider the *experimental philosopher* as conceived by Francis Bacon (1561–1626), often seen as the originator of the scientific method. Bacon argued that the systematic investigation of natural phenomena through experimentation would enable the collective human mastery of nature, a "kingdom of man" in which humanity would again attain the conditions of ease and harmony with nature enjoyed before the Fall (Noble, 1999). His utopian novel *New Atlantis* describes "Salomon's House", a college of learning with a range of facilities, including towers half a mile high where atmospheric phenomena can be closely observed, and "great and spacious houses" where they could create artificial snow, hail, rain, and lightning (Zagorin, 1998).

With this final character we arrive at a point that is similar to how we might imagine the modern geoengineer. In Bacon's vision we can see what sounds like a very well-resourced modern research institute. He describes something very close to modern science with its organised, collective character, its norms of objectivity and cooperation, and the pursuit of knowledge as an end in itself. Bacon also prefigures science's contemporary use for the common good in major civilisational projects like combatting disease and anthropogenic climate change.

However, let us not just discard the earlier characters outlined above. Note how they shift in complex ways back and forth between collective and individual benefit; between social, conversational models of weather control versus causal ones; and between emphases on keeping things stable or effecting change. And note also the following broad pattern that can be discerned as we move through the history of religion and culture towards modern times.

From harmony to dominance

Ancient societies typically understood social order and power in terms of *harmony* with the cosmos, and saw the aspiration to mastery over nature as dangerous and foolhardy. In such societies, over-reaching technical ambition was seen as disconnecting those who wield it from the wider sacred and social order, and likely to lead to disaster and punishment – as seen in the myths of Babel, Prometheus, and Icarus. Power came to those like the sacred kings who, instead, aligned themselves with the forces of nature. And this is a theme that is retained as it is transformed in our history above: even Bacon, with his emphasis on controlling nature and improving it, felt that "nature to be commanded must be obeyed" (Bacon, [1620] 1960, p. 39).

From inside to outside

Earlier understandings of the manipulation of the material world typically regarded the person who wields that power as an actor within the world – as an agent within a web of mutually interdependent agents. Even the Renaissance magi, who felt themselves to have access to secret knowledge about the occult properties of things, only thought they could alter the world if they operated within its organic order (Merchant, 1980, p. 169). With Bacon we can see a decisive shift: the modern understanding of mastery over nature is based on the idea of knowing the world from outside – of disentangling oneself from the world in order to gain objective and practically effective knowledge about its causal mechanisms.

From moral to technical action

In earlier societies the practical arts were also understood as bound up with morality – about the moral character of the person trying to wield power, and shared ideas about the good. Even the Renaissance magus's ability to manipulate the world is understood in social rather than purely technical terms, and thus to depend on his moral character. The first Renaissance magician, Marsilio Ficino (1433–1499) thought that to gain magical power the magus must enter into the "common love" that binds the world together, whereby "the lodestone attracts iron, amber straw, brimstone fire, the Sun draws flowers and leaves towards itself, the Moon the seas" (Easlea, 1980, p. 94). Bacon too linked knowledge and power to character; but for him nature would submit to the scientist if he purified his mind from emotion and disciplined his body, making it a suitable receptacle for the rational mind (Keller, 1985).

From active to passive matter

The ancients understood craft production and other forms of technical action as cooperating with a matter which had its own desires and goals; this notion of matter as active, even alive, shapes the thought and actions of our first three characters above. But Bacon advocated a "masculine birth of time" – a new start for the world's history, after which it would be dominated not by the unruly female principle of generation but by the male principle of rational control (Merchant, 1980, p. 168). Bolstered by the Reformation's insistence on God's utter transcendence from and divine mastery over his creation, this idea of matter as passive would help to lay the ground for science's aspiration to describe the world as a mechanism operating according to mathematical laws.

Situating climate geoengineering against this long, complex history suggests a number of questions. First, if we try to conceive what a "maker of climate" would be (Galarraga and Szerszynski, 2012), is our imagination too constrained by the contemporary secular mindset? Were historians to look back on current geoengineering debates from the vantage point of a few centuries hence, would they really conclude that we had fully understood the cultural currents and shifts that shape our thinking? Second, what difference might an awareness of religious and cultural history make to discussion about the ethics and governance of geoengineering? Can we learn from earlier ideas of harmonious and moral action, in ways that do not simply translate them into modern terms and lose their distinctiveness? Third, what relevance might this kind of inquiry have to the technical aspects of geoengineering? If the idea of geoengineering had emerged in a culture shaped by a very different trajectory, for example, one in which technical knowledge and action are seen as taking place within the world, and in which matter is understood as having its own agency, would it involve very different ways of thinking about how one alters or stabilises climate – ways which in the long term might be more productive?

First published online as an opinion article in 2014.

4

A HISTORY OF WEATHER AND CLIMATE CONTROL

James Rodger Fleming

Visionary schemes for weather and climate control have a long history, but with very few exceptions have ever worked. Would-be climate engineers and policy-makers need to take this into account. My intent here is to demonstrate that – contrary to claims that climate engineering is something wholly new in scale and intent – a number of previous technological interventions have been attempted on the atmosphere, on both regional and planetary scales. By and large, they did not have their desired effects on the physical environment, outpaced their original technical requirements, and gave rise to complicated political, social, and economic issues.

I would begin by addressing a claim that although historical cases of weather modification provide a valuable context for thinking about climatic interventions, they represent different temporal and spatial scales, and there-fore may be of limited comparative value. Manipulation of weather and cli-mate phenomena is intimately related. Any intervention in Earth's radiation or heat budget (such as managing solar radiation) would affect the hydrologi-cal cycle and the general circulation, thus rainfall and upper-level wind pat-terns, including the location of the jet stream and storm tracks. The weather itself would be changed by such manipulation. Conversely, intervening in severe storms by changing their intensity or their tracks or modifying weather on a scale as large as a region, a continent, or an ocean basin would obvi-ously affect cloudiness, temperature, and precipitation patterns, with major consequences for monsoonal flows and ultimately the general circulation. If repeated systematically, such interventions would influence the overall heat budget and the climate.

The earliest documented cases were rain-making schemes, and as such tended to be regional rather than global. In 1841 James Espy, America's first national meteorologist, developed a theory of storms powered by convection, but the so-called "Storm King" went off the deep end technically when he proposed lighting giant fires all along the Appalachian Mountains to emulate an artificial volcano that he thought would generate rains, disrupt cold and heat waves, and clear the air of miasmas (Espy, 1841). His contemporary, Eliza Leslie (1841),perceptively pointed out that attaining such control might cause serious damage to social relations. There were many other such rainmaking schemes (Leslie, 1842). In the 1920s, with concerns about aviation safety ascendant, independent inventor L. Francis Warren and Cornell chemistry professor Wilder D. Bancroft developed a scheme to dose the clouds with electrified sand delivered by aeroplane. Rainmaking and fog clearing were both on the agenda, but trials, supported by the US Army Air Corps, were less than promising. It turned out that aircraft could successfully disrupt smaller clouds, but experimenters could not predict whether a treated cloud would subsequently dissipate or thicken (Warren, 1925).

These early weather modification plans (some of surprisingly large scale) were couched in the context of the pressing issues and available technologies of their eras: Espy wanted to purify the air and make rain for the East Coast, and Warren and Bancroft hoped to make rain and clear airports of fog, while the military sought advantages for its fliers. But intervention is not control, and the hype surrounding both projects exceeded technical capabilities.

Prospects for larger-scale, even planetary intervention in the climate system arrived after 1945 with the dawn of several transformative technologies: nuclear weapons, digital computing, chemical cloud seeding techniques, and access to space (see Table 4.1). Two of the projects listed here involved cloud seeding techniques, and two involved disruptions of the space environment. All were part and parcel of the Cold War quest to militarise the atmosphere. Not listed in the table are proposals, dating from 1945, to bomb nascent hurricanes or break up polar ice with nuclear weapons, or to build a digital computer that would produce perfect forecasts and perhaps allow real-time intervention in threatening weather systems as they developed.

In 1947 scientists at the General Electric Corporation developed methods for seeding clouds with dry ice and silver iodide, sparking a race for commercial applications and military control of the clouds. They partnered with the military in Project Cirrus to seed an Atlantic hurricane with dry ice, but the experiment went awry. Nevertheless, GE chief scientist Irving Langmuir (1948) hyped the possibilities, arguing that hurricanes could be redirected and that the climate

TABLE 4.1 Weather and climate control projects in the Cold War

1947	Project Cirrus attempts diversion of an Atlantic hurricane using dry ice seeding.
1958	Project Argus, top-secret military project detonates three atomic bombs in space.
1962	Starfish Prime, H-Bomb detonated in magnetosphere. Similar Soviet tests.
1967	Monsoonal cloud seeding over Vietnam leads to UN ENMOD treaty in 1978.

Source: Fleming, 2010.

might ultimately be controlled on a continental or oceanic scale with the techniques they had developed. Cloud seeding reached around the world, especially into arid areas and upslope watersheds, but they never resulted in fully reliable techniques to enhance precipitation or snowpack. The scale of nature was too huge and problems of verification and social acceptance were also too huge. Instead of quasi-military aerial bombardment of the clouds, small-scale practices such as drip irrigation and snowmaking machines became the norm.

Between 1966 and 1974 massive and surreptitious seeding of the South-East Asian monsoon during the Vietnam War resulted in little measurable rain, but a diplomatic nightmare for the United States when the Soviet Union brought the issue of environmental warfare to the attention of the United Nations. The UN Convention on the Prohibition of Military or Any Other Hostile Use of Environmental Modification Techniques (ENMOD) was the biggest fallout from the effort, followed by a systematic and persistent collapse of US federal support for cloud seeding (Fleming, 2006; United Nations, 1978).

The Argus and Starfish Prime nuclear detonations in space, along with similar Soviet testing, constituted actual attempts to engineer space weather and disrupt the magnetosphere. A theory promulgated by Nicholas Christofilos, a physicist at Lawrence Berkeley Lab, held that the ionised debris and high-energy electrons generated by a nuclear explosion would travel almost instantly through Earth's magnetic field as a giant current. In case of hostilities a nuclear blast could possibly generate a massive electromagnetic pulse over an enemy city, disrupt military communications, and destroy both satellites and the electronic guidance systems of enemy missiles. These tests, conducted by both superpowers, generated widespread public outrage and were quickly followed by the Limited Test Ban Treaty (Christofilos, 1959; Fleming, 2011).

Lessons from history for weather and climate engineering

History teaches us that things change – often in surprising or unanticipated ways – and that a certain amount of clarity can be gained by looking backward as we inevitably rush forward. Schemes aimed at attempted control of weather and climate – often framed as responses to critical problems such as water shortages, military exigencies, and cold war dominance – have fallen short of their goals many times in the past. The chequered history of this field provides valuable perspectives and a cautionary warning on what might otherwise seem to be today's completely unprecedented climate challenges. Contemporary engineers err if they ignore this history.

Would-be climate engineers are strongly motivated by fears of future global warming, but within recent memory this landscape too has been changing. Our scientific understanding of climate sensitivity to greenhouse gases remains roughly as uncertain today as it was two decades ago, showing the complexity of the climate system geoengineering proposes to modify. Additionally, there is strong technical resistance, or at least caution, from the faculty of mainstream atmospheric science departments, who tend to be sceptical of simple geoengineering schemes. Increasingly, historians, philosophers, and other humanists and social scientists are getting beyond back-of-the-envelope technicalities and are taking a critical look at complex issues related to the history, ethics, and governance of global control issues. Even the neologism "geoengineering" is in the process of being abandoned (since it is not really engineering in any traditional sense), as is the phrase "solar radiation management" (since there are too many unknowns to really consider it a form of management).

Intervention into weather and climate systems does not result in control over them. Instead it has often given rise to unexpectedly complicated social issues. We should base our decision-making not only on technical expertise and what we think we can do "now" and in the near future. Rather our knowledge must be shaped (and tempered) by what we have and have not done in the past. Such are the grounds for making informed decisions and avoiding the pitfalls of rushing forward claiming we know how to control weather and climate.

As a concluding reminder of the challenges and importance of considering the historic backdrop to contemporary discussions of geoengineering, the following misleading claims were made by various speakers at the 2010 Asilomar International Conference on Climate Intervention Technologies, with my brief responses provided in italics:

> "We don't have a history of geoengineering to fall back on …" – *Yes we do.*

> "Things are moving quickly, so we don't have the luxury of looking at history". – *We must take the time.*

> "We are the first generation to think about these things". – *History says otherwise.*

First published online as an opinion article in 2013.

5

NITROGEN GEOENGINEERING

Oliver Morton

Discussions of climate geoengineering often treat it as an unprecedented venture. James Fleming's perspective in the preceding piece as well as his previous writing (2010) on the history of weather modification clearly show that this is not the case; people have been discussing manipulations of weather and climate, and developing technologies to that end, for a long time. Acknowledging this provides perspective and context for discussing proposals that are current today.

The history of geoengineering, though, is not to be found entirely in the history of climate studies. If geoengineering is taken to be the large-scale and purposeful technological manipulation of the Earth system to a given end then the development and deployment of industrial nitrogen fixation provides an informative precedent.

The 20th century's development of nitrogen geoengineering is the most dramatic example history provides of humans changing the way the Earth system works. Those interested in bringing about a conceptually similar, if in some ways more modest, change through climate geoengineering would do well to familiarise themselves with this precedent's inception, its scale of action, and its consequences, even while keeping in mind the limited value of historical analogues in such matters.

The technology required for the industrial takeover of the nitrogen cycle did not appear through an unguided process of innovation, nor was it deployed that way; the foresight involved is part of what makes it a geoengineering technology in a way that other agricultural innovations, and indeed agriculture itself, are not. Nitrogen fixation was developed purposefully in response to a threat, which, while not obvious in everyday life, had been identified by the scientific elite. Like climate change today, that threat was seen as being of global significance and to have no easily attainable political solution. That justified a concerted effort to develop a technological response. Though people working in the climate arena may not immediately recognise this response as an example of geoengineering, some of those working on the nitrogen cycle have no problem seeing it as such (Sutton et al., 2011).

Fixing the nitrogen cycle

The nitrogen cycle, no less than the carbon cycle, is fundamental to the way the Earth's biosphere works. The bacteria that "fix" nitrogen from its inert gaseous form into compounds that can be made use of by plants, and the animals that eat them, are a key part of this cycle. Though some inorganic processes fix nitrogen, bacteria in the soil and seas are the overwhelmingly dominant natural source of such compounds.

Crop yields can often be increased by the addition of extra fixed nitrogen. Historically this was achieved at various times and places by treating the soil in ways that increase bacterial nitrogen fixation or by importing nitrogen in the form of manures. In 19th-century Europe these techniques were supplemented with the use of nitrate minerals—resources that were also needed by the chemicals industry, notably for the production of explosives. In the 1870s strategic control of the richest South American nitrate deposits was the principal cause of the "War of the Pacific" (also called the Saltpetre War) between Chile, Peru, and Bolivia.

Some men of science became aware that the world's supply of nitrate was insufficient to cope with its remorselessly rising demand for food. Noting the ever-greater demand for wheat and the lack of new land on which to grow it, Sir William Crookes, a noted British chemist, used his 1898 presidential address to the British Association for the Advancement of Science to stress that it was "vital to the progress of civilized humanity" that chemists solve the problem by fixing nitrogen from the air into compounds that could be used as fertilisers and chemical feedstocks. If they failed there would be a planet-wide "catastrophe little short of starvation ... and even the extinction of gunpowder!". In 1908 Fritz Haber hit on a successful scheme; Carl Bosch made its use a practical industrial process, and Germany's need for explosives in the First World War saw the process put into large-scale use (Smil, 2004). By the 1920s the Haber-Bosch process – subjected to compulsory licensing as part of the Treaty of Versailles – was available to industries throughout the world and chemistry textbooks were congratulating themselves on having solved the "nitrogen problem".

It has been estimated that the explosives made possible by the Haber-Bosch process contributed directly to 150 million deaths over the 20th century (Erisman et al., 2008). In the second half of the century, though, it made an even greater contribution to life than it had to death. Nitrogen-based fertilisers were the single greatest contributor to near tripling of crop yields in the decades after 1950. Today fertilisers produced by the Haber-Bosch process account for almost half of the nitrogen in human food; without them the population would not have been able to grow close to its present seven billion. By the time the population stabilises somewhere around ten billion, most of the nitrogen in those peoples' muscle fibres, nerve cells, and DNA will be coming from factories.

Industry now far outdoes the world's soil bacteria in the fixation of nitrogen. This makes human intervention in the nitrogen cycle considerably greater, in proportion to the natural flows, than is the climate-changing human intervention in the carbon cycle.

Lessons for climate geoengineering

In what ways can this historical analogue inform debates about climate geoengineering? First, it offers an existence proof. It is possible for humans to identify a global problem, create a technology that addresses that problem, and deploy it on a global scale.

It also shows that the scope of such an intervention can greatly outstrip its progenitors' plans, and perhaps their imaginations. While Crookes did not put specific numbers on the amount of nitrogen he felt was needed, the scale of today's nitrogen industry and its effects surely far outstrip the consequences he expected. This suggests that those imagining possible futures for climate geoengineering should take care that they imagine applications of the technology well beyond the minimum that seems to be required – not as necessary endpoints to the programme, but as plausible ones.

There are other aspects of nitrogen geoengineering that climate geoengineers should be aware of. One is that it is deployed inefficiently. Most of the deliberately fixed nitrogen does not get into crops; Vaclav Smil estimates that the overall efficiency of the global food system in terms of nitrogen use is less than 15 per cent (Smil, 2011). The wasted nitrogen is not just a loss; it often does harm. Over-fertilised soils produce nitrous oxide, which destroys stratospheric ozone and is also a powerful greenhouse gas. Nitrate-bearing run-off waters from agricultural watersheds stimulate algal blooms and "dead zones" in coastal seas. While nitrogen fixation has made the world more habitable by humans – more precisely, habitable by more humans – than it otherwise would have been, it has also done significant damage to biodiversity, human health, and ecosystem services in the process.

The waste problems are made more complicated by the fact that humans fix nitrogen inadvertently as well as deliberately. The nitrates produced as by-products of combustion in vehicles and industrial plant do a great deal of harm, most notably through particulate emissions which damage human health. They also stimulate the growth of unfertilised ecosystems such as European forests and, indeed, organic farms. The effects of deliberately fixed nitrogen and inadvertently fixed nitrogen blend into each other in various ways.

There is a general lesson here; the side effects of geoengineering will often be intermingled with the effects of inadvertent pollution involving related substances. This does not make geoengineering indistinguishable from pollution. There is a qualitative, indeed categorical, difference between geoengineering to a specific end and heedlessly making a planet-sized mess. At the same time, nitrogen geoengineering shows that, when inspected closely, the dividing line between deliberate action and pollution can be disturbingly blurry.

Many other parallels (and distinctions) might suggest themselves. I will close on two. One is that geoengineering is surprisingly easy to overlook. To its proponents and opponents, climate geoengineering currently seems a fundamental and historic transition. After the fact it might look much less vexatious. The billions benefiting from nitrogen geoengineering hardly know it is going on. The same might well be true of, say, an aerosol layer in the stratosphere that reduced incoming solar radiation by a watt per square meter. Many people would know it was there and some would oppose it (it is worth remembering that the organic farming movement began in large part as a response to industrial nitrogen fixation). But it might well be only a minor concern to most.

The second closing point is to draw out a difference. While there was a well-articulated need for nitrogen fixation that drove the development of the technology, its deployment was often decentralised. This was not always the case. In China, and other smaller economies, the rearrangement of the nitrogen cycle was planned at a national level. The deployment of fixed nitrogen was fundamental to the political goals of the "green revolution" and was coordinated accordingly. In this nitrogen geoengineering was quite distinct from agriculture-as-usual. But the dynamic of the technology's spread was shaped by the fact that local and regional benefits could be achieved independently of the global picture, and their benefits captured immediately and privately.

Some climate geoengineering approaches might spread in a similar way: crop-albedo changes, perhaps, or cloud brightening aimed at steadying monsoons. But such a piecemeal dynamic is not likely to shape climate geoengineering techniques with primarily global impacts, such as stratospheric aerosol injections. This all-or-nothing attribute of climate geoengineering may well make it harder to achieve than nitrogen geoengineering was. But it might also offer the possibility of a better-designed intervention—its benefits optimised, its intentionality well governed, its burdens shared equitably.

First published online as an opinion article in 2013.

PART II

Contemporary framings

Introduction

Crutzen's reframing of geoengineering in his 2006 article – as perhaps being able to forestall or ameliorate the future ravages of climate change, should humanity not act quickly enough to avoid them – attracted unprecedented new attention to climate engineering concepts (Crutzen, 2006). While potentially a compelling narrative for those concerned that current governments are unlikely to muster the political will to sufficiently mitigate climate change, this framing is fraught with unanswered questions – questions that the essays in this section of the book unpack and explore in detail.

Tim Lenton opens this discussion in Chapter 6, examining the scientific validity of the notion that geoengineering technologies could be effectively held back and deployed at the last minute to avoid "climate emergencies". His analysis leads to the challenging conclusion that perhaps only "pre-emptive geoengineering" interventions could effectively avoid the risk of some dangerous climate tipping points. In Chapter 7, economist Juan Moreno-Cruz and colleagues examine the "fast, cheap, and imperfect" nature of proposals to use stratospheric aerosols to cool the Earth, referencing (and providing an accessible summary of) economic game theoretical studies that model the strategic actions of states in deploying them, and that thereby frame potential geopolitics. Based on the low projected cost[1] of deploying stratospheric aerosols, they argue that a purely economic framing for geoengineering could lead to undesirable outcomes that need to be governed against.

In Chapter 8, Rob Bellamy steps back to examine how internationally respected scientific assessments have framed geoengineering technologies over the last decade. He particularly notes the gaps created by these assessments focusing only on geoengineering in isolation from other responses to climate change, and the risk of some narratives getting prematurely locked-in and short-changing the public's opportunity to have an open debate. In Chapter 9, Rose Cairns continues with this theme, reflecting on the challenges of coming to understand what the public – or different "publics" – thinks about an idea as complex as geoengineering, particularly when the technologies are still largely imaginary. In Chapter 10, Jeff Tollefson takes a practitioner's perspective, reflecting on the challenges of communicating about geoengineering to the public as a science journalist, and the difficulty of reducing "complexity and nuance, to headlines and column inches". These contributions reflect upon the tensions faced by researchers – or public-facing communicators, more broadly – in marshalling knowledge and framing debate for deliberation by wider audiences. Certainly, engineering and modelling assessments have influential framing implications, which stakeholder engagements led by social scientists have sought to engage and democratise – yet all researchers need to be watchful of their own effects in structuring public discourse (Bellamy and Lezaun, 2017). This call for reflexive research links to the underpinning ideas of "anticipatory" frameworks for governing novel and controversial technologies, which we turn to in our governance section (Part VI, as well as Foley et al. in this volume).

The last two chapters of this section return to broad themes raised in Part I; specifically, the religious and ethical significance of different ways of framing geoengineering. In Chapter 11, Wylie Carr presents his own research demonstrating that religious and cultural beliefs can provide a pre-existing frame which religious people will draw upon to interpret geoengineering ideas, reminding readers of the perception that "This is God's stuff [geoengineering is] messing with". Finally, in Chapter 12, Nancy Tuana uses the example of stratospheric aerosol injection (SAI) to identify a comprehensive series of questions that push beyond simple framings of geoengineering to explore the underlying ethical challenges and nuances of these planetary-scale technologies. This kind of technology-specific research agenda, seeking to couple ethical and scientific analyses for use in concrete policy and governance questions, can be seen as an example of the "second wave" of approach-specific, context-driven ethical studies referred to in Baatz et al. (2016). However, it also shows that SAI raises classic questions with respect to intergenerational, distributive, corrective, ecological, and procedural justice in new ways. A recent updating and summarising of these ideas in light of SRM approaches can be found in Svoboda (2017).

The contributions of this section reflect upon the construction of frames currently exercised upon the viability and desirability of geoengineering: how they are generated, by whom, for whom, and with what potential influences. It does not map the frames themselves. Here we might consult a number of discourse analyses and bibliometric studies that map the economy of frames, narratives, moral arguments, and metaphors – as embedded in academic/policy literature and popular media – across stakeholder groups or national contexts. Since these are not reviewed in any of the book's contributions, we list much of the literature below, which reflects the strong shaping influences of academic networks based in North America and Europe from 2006 onward. This literature is an incomplete survey of constituencies and perspectives in the geoengineering space, with fewer studies that solicit and map public and policy perspectives compared to academic and media framings. We turn to a further set of national, regional, and sectoral perspectives in our Part V.

For now, we note some broad trends from the literature. An increasingly dominant theme emphasises that geoengineering is a "risk-risk" proposition, that must navigate tension between its own risks and those of insufficient mitigation and adaptation measures (Huttunen and Hilden, 2013; Huttunen et al., 2014; Linner and Wibeck, 2015). A more enthusiastic frame calls for an ideological change of tack from conservationism, in which complex human-environmental systems – such as the climate – need to be proactively intervened in and managed (Anshelm and Hansson, 2014a, 2014b; Buck, 2013; Scholte et al., 2013; Cairns and Stirling, 2014). Critical perspectives emerge as well: many warn of the dangers of applying "technofixes" to complex systems, of perpetuating the carbon economy and the supposed geopolitical predominance of the global North in geopolitics, or of "leaving science to scientists" in issues of moral and political import (Bellamy et al., 2013; Corner et al., 2013; Macnaghten and Szerszynski, 2013).

Note

1 A further dimension worth noting is that the "cheap" costs of solar geoengineering have been questioned and revised since the original engineering assessment that underpins that trope (McClellan et al., 2011). Moriyama et al.'s (2016) more recent assessment of technical implementation costs revises the numbers upward; MacKerron (2014) provides wider criticisms that bracketing cost calculations to implementation elides political and economic effects, and that large capital projects often suffer from "appraisal optimism", or vast budgetary underestimations. These observations, however, do not fundamentally erode Moreno-Cruz et al.'s (2011) points.

6

CAN EMERGENCY GEOENGINEERING REALLY PREVENT CLIMATE TIPPING POINTS?

Timothy M. Lenton

One broad framing of geoengineering is that it could be used to try to avoid a "climate emergency". Several types of "climate emergency" have been implied in the literature, ranging from abrupt, non-linear changes in the climate system itself – referred to here as "climate tipping points" (Lenton et al., 2008) – to sudden changes in social systems that might be triggered even by gradual climate change – referred to here as "impacts tipping points".

In particular, it has been argued that potent methods of sunlight reflection, such as stratospheric sulphate aerosol injection, might be deployed in an emergency if a part of the Earth system is seen to be approaching, or to have passed, a climate tipping point. For example, a report from the Bipartisan Policy Center considers "most climate remediation concepts" to be "inappropriate to pursue except as complementary or emergency measures – for example, if the climate system reaches a 'tipping point' and swift remedial action is required" (BPC, 2012). This begs the question: by the time you realise you have reached a tipping point, can you actually go back?

The Royal Society offers a subtly different framing of "Solar Radiation Management techniques" that "because they act quickly, they could be useful in an emergency, for example to avoid reaching a climate 'tipping point'" (Shepherd et al., 2009). This begs a different question: how do you know you are approaching a tipping point? And if an early warning is possible: can you find out early enough and act fast enough to avoid reaching a tipping point?

The fundamental problems with the "emergency-use" framing of geoengineering are that the parts of the climate system which may pass a tipping point are lagging behind anthropogenic forcing, and that passing a tipping point can lead to irreversible change. So, by the time you detect that abrupt change is either imminent or underway, the tipping point may long since have been passed and the change simply cannot be reversed.

To illustrate irreversibility, consider the special case of a tipping point that is a "fold" bifurcation in the equilibrium solutions of a system (Figure 6.1). This could be the Atlantic Ocean's overturning circulation (Stommel, 1961; Rahmstorf, 1995) or a large ice sheet such as on Greenland (Gregory et al., 2004; Ridley et al., 2009). When it reaches a tipping point, the current state of the system loses its stability and it undergoes an abrupt transition to an alternative state. The timescale of this transition is set by the internal dynamics of the system in question and can range from years (e.g. past abrupt warming events linked to reorganisations of the Atlantic overturning circulation) to centuries (e.g. for the melt of large ice sheets). Once in this alternative state, the system has to be taken to a different and distant tipping point to trigger recovery. Even after that, it is not back where it started.

Irreversibility is even stronger in the case of tipping points in ecological systems, for example dieback of the Amazon or boreal forests (Lenton et al., 2008). If species become extinct that is final, and particular configurations of ecosystems may also be unique and unrecoverable. Impacts tipping points in social systems may also be irreversible, for example, if a low-lying coastal city is abandoned in response to steady but overwhelming sea level rise. Putting the coupling between ecosystems and human systems into the equation probably only adds to the irreversibility, for example, when an established agricultural system becomes unviable and this triggers social unrest or mass migration.

Now add the problem of lag (Figure 6.1b); sluggish systems, such as the ocean circulation and ice sheets, cannot keep up with the rate of anthropogenic climate change. This means they are no longer in equilibrium with the climate forcing and instead they are in a transient state, lagging behind it. Even ecosystems such as the Amazon rainforest may lag climate forcing by several decades (Jones et al., 2009). When they start to show signs of abrupt change, they will already have overshot a tipping point and are well into the "basin of attraction" of an alternative state. Any geoengineering applied at this intervention point has to fight against the system's own dynamics which are trying to take it into an alternative state.

If this all sounds a bit technical, picture Looney Tunes' Wile E. Coyote chasing Roadrunner to the edge of a cliff but overshooting spectacularly – he hovers tantalisingly in the air, far above the ground below (the lagged response), but his fate is already sealed. Geoengineering to alter his fate (in this metaphor) amounts to trying to propel Wile E. back onto the cliff top before gravity takes him abruptly to the alternative state on the ground below.

Thus, if one waits to see abrupt climate change unfurling, we may long since have been committed to it, and "swift remedial action" (BPC, 2012) is actually not swift at all, however quickly geoengineering can be deployed and take effect.

It may still be possible in principle to reverse a change that is underway, but it could demand a reduction in radiative forcing well below the pre-industrial level (Figure 6.1a). For example, the Greenland ice sheet is thought to be a relic of the last ice age – if it is removed it will not regrow under the pre-industrial climate (Toniazzo et al., 2004). To reinstate it would require us to geoengineer a much cooler climate that would be undesirable for many other reasons.

Those suggesting emergency geoengineering as a remedial action once a tipping point is passed need to focus their attention on tipping points that have the greatest reversibility, for example, abrupt loss of summer Arctic sea-ice cover (Tietsche et al., 2011).

As for geoengineering to "avoid reaching a climate 'tipping point'" (Royal Society, 2009), there is a glimmer of hope in that systems approaching bifurcations carry generic early warning signals, such as becoming more sluggish in their recovery from natural fluctuation (Lenton, 2011a; Scheffer et al., 2009). These warning signals were present prior to some abrupt climate changes in the palaeo-record (Livina and Lenton, 2007; Dakos et al., 2008), and are also found in models being slowly forced towards tipping points (Held and Kleinen, 2004; Lenton et al., 2009). However, the lag problem is still pertinent; we are forcing many parts of the climate system so rapidly, relative to their internal dynamics, that warning signals may not be detectable in advance of reaching a tipping point.

So, those suggesting geoengineering to avoid reaching a tipping point would do well to focus their attention on fast-responding systems which should carry the best early warning prospects (Lenton, 2011a), for example, monsoons or (once again) the Arctic sea-ice. Interestingly, these may also be among the more reversible systems. However, monsoons are particularly sensitive to aerosol forcing and may actually be disrupted rather than protected by deliberate aerosol injections (Ramanathan et al., 2005; Lenton, 2011b).

One could take a wider view of "early warning" and note that threats from multiple tipping points are already recognised, and one does not need a direct early warning signal to act. "Pre-emptive" geoengineering might then be considered as a means of avoiding what are thought to be dangerous magnitudes, rates, or gradients of climate change (Lenton, 2011b). But in that case, it seems questionable that the geopolitical willingness could be generated to geoengineer on a pre-emptive basis. After all, we have not yet succeeded in pre-emptively (or retrospectively) reducing emissions of carbon dioxide and other warming agents.

Part of the reason for this failure to act may be uncertainty over the proximity of tipping points. There is experimental evidence to suggest that the fear of crossing a dangerous threshold could turn climate negotiations into a coordination game, making collective mitigation action to try to avoid the threshold virtually assured. But current uncertainty about the location of tipping points instead causes cooperative efforts to avoid dangerous climate change to fail (Barrett and Dannenberg, 2012).

The irony here is that the surest way to reduce uncertainty about the location of climate tipping points is to get closer to them, but by then effective cooperative action to mitigate greenhouse gases emissions is likely to be too late to avoid tipping points. Instead sunlight reflection methods might be the only viable

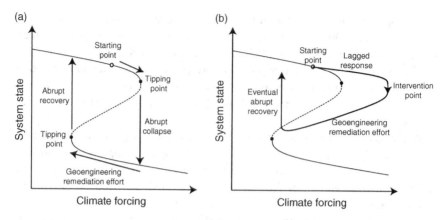

FIGURE 6.1 A tipping point that is a "fold" bifurcation in the equilibrium solutions of a system.

avoidance option left on the table, and they are only viable for a subset of tipping points that are directly related to temperature change (Lenton et al., 2008).

Whether adding pre-emptive geoengineering into the mix of policy options will help achieve cooperative action on climate change is unclear. Our own experimental economics research suggests that the presence of multiple options, such as mitigation and geoengineering, can contribute to inefficient failure to coordinate collective action to avoid dangerous climate change.

In summary, "emergency-deployment" framings of geoengineering to avoid or reverse climate tipping points could be seriously flawed. It needs to be researched under what – if any – circumstances geoengineering could actually work to avoid a tipping point or to reverse one that had been passed. While this might be taken as an argument for "pre-emptive" geoengineering to reduce the risk of reaching a tipping point, it is not at all clear that introducing this option into the policy mix will actually help trigger effective action to avoid dangerous climate change.

First published online as an opinion article in 2013.

7

THE ECONOMICS OF CLIMATE ENGINEERING

Juan B. Moreno-Cruz, Katharine L. Ricke,
and Gernot Wagner

Fast, cheap, and imperfect

Unmitigated climate change is extremely costly. Mitigation (the reduction of carbon dioxide and other greenhouse gas emissions at the source) is the only prudent response. While upfront costs can be high, in the long run mitigation is relatively cheap and because it tackles the root cause of the problem, its benefits are permanent and transparent. However, because the effects of mitigation investments are subject to considerable physical and social inertia, they are slow to manifest. Moreover, effective mitigation requires overcoming the well-known "free rider" effect inherent in the most global of global commons problems.

Enter climate engineering in the form of planetary albedo modification, or solar radiation management (SRM). Although it is both cheap and fast, SRM is clearly an imperfect method of countering climate change. Rather than addressing global warming's root cause, it counteracts its effects with additional pollution. But SRM is so cheap that direct private costs largely do not play a role in the decision process of whether (and how) to pursue it (Barrett, 2008, 2014; Moreno-Cruz and Keith, 2012; Keith, 2013; Wagner and Weitzman, 2015). Well-designed SRM systems could cost less than $10 billion per year to stabilise global temperatures. That places implementation within the budget of many countries, making SRM perhaps significantly cheaper than any emissions reduction programme to stabilise global average temperatures below a post-industrial increase of 2°C (Morgan and Ricke, 2010; McClellan et al., 2010; Moriyama et al., 2016). It is so fast that it can reduce planetary temperature in days, weeks, or months, compared to the decades or centuries that emissions reductions would take to have a similar effect.

"Fast, cheap, and imperfect" (Keith et al., 2010) give SRM virtually the exact opposite economic properties from mitigation. Instead of the classic free-rider effect in which actors are compelled to underprovide mitigation, the chief economic characteristic comes much closer to a "free driver" effect in which actors are compelled to overprovide SRM (Weitzman, 2015; Wagner and Weitzman, 2015). In terms of policy interventions, this means that rather than pursuing individual actors to mitigate counter to their immediate personal interest, the task becomes one of reining in those who want to pursue climate engineering of their own volition.

Economic tools to analyse SRM

Two types of standard economic tools are most prominent in the economic analysis of geoengineering: benefit-cost analysis and game theory.

Benefit-cost analysis attempts to weigh the full set of benefits and costs of any policy intervention (in this case, SRM) to make informed decisions. To identify the optimal policy, researchers make assumptions about social benefits and costs associated with both SRM and mitigation and the damages associated with unmitigated climate change. The optimal policy balances the benefits of action against the costs associated with the action taken.

Well-known problems exist in the application of conventional benefit-cost analysis to problems of global-scale change (Morgan et al., 1999). One inherent problem with such analyses is their reliance on an assumption of modest and quantifiable uncertainty (Goes et al., 2011). In the case of SRM, estimates of both climate change's damages and the costs of geoengineering fail to meet this criterion. While we have a relatively good handle on the direct, private costs associated with SRM, the full sets of social costs – and the full social benefits, for that matter – are largely a mystery (Moreno-Cruz and Keith, 2012; Wagner and Weitzman, 2015).

Perhaps an even more fundamental problem with these kinds of benefit-cost analyses is that they largely assume optimally coordinated decisions across regions or countries. In particular, they ask: what are the total benefits and costs of SRM from a benevolent, global social planner's perspective seeking to optimise policy for the world? Reality, of course, does not conform to this assumption.

This is where game theoretic and political economy studies come to play. Game theory provides a tool to reveal frictions that arise when self-interest actors wish to maximise their own well-being without considering the full effects of their actions on others. Instead of optimising outcomes from the perspective of a single decision-maker, in a game theoretic model multiple players interact and make strategic decisions to optimise their own individual outcomes.

The primary limitation of game theoretic analyses of SRM to date may be a lack of any consensus over the key constraints that would govern strategic interactions in a geoengineered world. Different assumptions about how physical effects of SRM translate into economic damages, what the institutional characteristics of a sustainable geoengineering implementation programme would be or even who could feasibly get away with setting the global thermostat without consent from other world powers, lead to drastically different conclusions about what outcomes are possible (Ricke et al., 2013; Millard-Ball, 2012; Moreno-Cruz and Keith, 2012; Urpelainen, 2012). The sensitivity of the outcomes predicted by game theoretic analyses to their model constraints implies that as uncertain and unpredictable as the science of SRM may be, political dynamics likely play an even more important factor in shaping the final outcome.

Rogue actor, rational portfolio, and worst-case insurance

Current international legal standards do not explicitly restrict any nation from engaging in stratospheric SRM (Parson and Ernst, 2013; Reynolds, 2015). It is unlikely that even if they did, such restrictions could effectively block action, just as the Nuclear Non-Proliferation Treaty did not prevent non-signatories like India, Israel, North Korea, and Pakistan from developing nuclear weapons.

Both legally and technically, a single nation or even a wealthy individual could take matters into their own hands. Unilateral SRM deployment could lead to the alleviation of climate change impacts for one group while imposing a mix of externalities on another group without its consent or compensation. Underestimating the full benefits and especially costs may only be a small issue in this situation. Full-on "climate wars" – with geoengineering and counter-geoengineering efforts – would be quite another. This is one of the reasons it is important to understand the potential harm and benefits associated with SRM, not only globally, but also in terms of relative regional effects, and important for responsible actors to develop governing mechanisms for careful, deliberate geoengineering research.

A deliberate approach to SRM may also link into two other potential geoengineering scenarios: SRM as part of a portfolio of strategies to address climate change, or as a deliberate backstop technological response to a global climatic emergency. The two are not mutually exclusive. They also come with their individual challenges.

As a part of a portfolio of strategies, together with mitigation and adaptation, SRM could in theory be used to achieve an "optimal" response to climate change, minimising net social cost or maximising some other global, societal objective (Moreno-Cruz et al., 2011; Moreno-Cruz and Keith, 2012). This framing is controversial for a host of reasons. For one, fundamental uncertainties of the underlying climate and ecological science may make a deliberate, deterministic choice difficult or impossible. Another reason is the host of ethical issues involved in intentionally engineering the planet. Moreover, the decision process is murky. There is no single global decision-maker for whom an optimum can be defined. Solving any of these challenges is difficult, to say the least.

Political economics could conceive of rational voting mechanisms that do define a globally optimal solution, though the practical difficulties of implementing any such scheme would be daunting (Weitzman, 2015; Wagner and Weitzman, 2015). Alternatively, countries could engage in negotiations to create a coalition with enough power to legitimise the "free driver" intervention while excluding a majority of countries (Ricke et al., 2013).

A perhaps less controversial argument is that SRM could serve as a form of insurance of last resort, in case of a worst-case climatic scenario. Likely climate outcomes should provide sufficient impetus for action on mitigation. But it may be the low-probability, high-impact events that dwarf any and all currently projected, average costs. A conservative calibration of what is known about climate sensitivity – the effect on eventual global average temperatures as greenhouse gas concentrations in the atmosphere rise – indicates that we are well underway towards a world with a greater than 10 per cent chance of global average temperatures rising above 6°C (Rogelj et al., 2012; Wagner and Weitzman, 2015). Such a scenario may well trigger a host of irreversible tipping points, many of which could occur much closer to the vaunted 2°C threshold (Lenton et al., 2008).

Such tipping elements include the disappearance of Arctic summer sea-ice (a process well underway), rapid decay of the Greenland ice sheet, or the collapse of the ocean's thermohaline circulation. It's clear that the potentially dire economic consequences of such phenomena contribute disproportionately to the risks associated with global warming. The extent to which SRM could stop or slow the progression of extreme climate change after its commencement or detection is unknown (Lenton, Chapter 6). That makes seeking insurance against worst-case scenarios a primary reason cited for investing in SRM research (Morgan and Ricke, 2010; Shepherd et al., 2009).

The insurance proposal does not come without drawbacks. For example, the technology would be developed by the current generation and would be implemented by a future generation, resulting in time inconsistency and sub-optimal use of the technology by future generations (Goeschl et al., 2013). Nonetheless, the downside risks of unmitigated climate change might be so large that virtually any intervention might pass a benefit-cost test.

From incredible to seemingly inevitable?

Perhaps the most apt description of the economics of geoengineering came in Barrett (2008), which described it as "incredible" due to the small costs of SRM. In fact, the costs are so small, they turn the standard economics of climate change as a global commons problem on its head – from the "free rider" to the "free driver" problem.

The very same fundamental market forces that make mitigation action so hard seem to point directly towards an SRM-style intervention. None of this should be misconstrued as an endorsement of this path. Quite the opposite: it's a call for a deliberate step towards a governing mechanism that could guide these market forces in the right direction (Honegger et al., 2013; Zürn and Schäfer, 2013).

First published online as an opinion article in 2015.

8

FRAMING GEOENGINEERING ASSESSMENTS

Rob Bellamy

The ways in which social actors choose to organise and communicate, or "frame", ideas such as geoengineering channel expressions of power. In support of decision-making, the selection of contexts and methods in technology assessment both constitute broad sites of instrumental framing, which act to condition assessment outcomes (Stirling, 2008). As an "upstream" suite of technology proposals that is currently in advance of significant research and development or public controversy (Wilsdon and Willis, 2004), geoengineering is particularly sensitive to these, and other, framings.

Assessments of geoengineering have so far largely taken place under two dominant problem definitions (Bellamy et al., 2012). First, that efforts to reduce greenhouse gas emissions will not be enough to tackle climate change ("insufficient mitigation"). Second, that as a result of this, we may be faced with a dangerous change in climate, often stylised as crossing one or more "tipping points" – a "climate emergency" (Lenton, Chapter 6). Both of these framings posit a central role for geoengineering in tackling climate change, concurrently marginalising legitimate alternatives. Accordingly, geoengineering assessments have largely excluded mitigation options and adaptation, placing the proposals in what can be termed "contextual isolation".

Although geoengineering framings have begun to emerge and diversify in the media (Porter and Hulme, 2013; Scholte et al., 2013) – with coverage likely to increase following its heightened prominence within the Fifth Assessment Report of the Intergovernmental Panel on Climate Change (2014) – it typically remains a scientifically and technically framed issue within the assessment literature (Bellamy et al., 2012). Most of the assessments use technical and analytic methods that are exclusive to experts (such as climate models or cost-benefit analysis), and narrowly focused, technical criteria (such as global temperature reduction, rapidity of effect, and cost). Under such methods and criteria certain issues are privileged, and so too are certain proposals. It is perhaps no surprise then that an ostensibly effective, fast-acting, and cheap proposal, stratospheric aerosol injection, often performs very highly and is gaining attention (Lenton and Vaughan, 2009; Shepherd et al., 2009; Bickel and Lane, 2009).

The expert multi-criteria assessment conducted for the Royal Society's seminal report provides a valuable illustration of how framing geoengineering assessment through the selection and elevation of particular criteria can compel particular outcomes (Shepherd et al., 2009). The assessment utilised four technical criteria with which to evaluate the proposals: effectiveness, affordability, timeliness, and safety. In then presenting the performances of the different proposals on a two axis figure, a difficult decision was made with respect to which of the four criteria would be given priority on those axes. It was decided that effectiveness and affordability would be given that normative priority, and under that configuration stratospheric aerosol injection performed the highest overall.

The prioritisation of effectiveness and affordability on those axes, however, was only one of a possible six permutations (Figure 8.1). Each of the differently framed permutations offers a distinct pattern of performances, where different overall conclusions can be drawn. Where the original configuration places stratospheric aerosol injection most highly (a), that performance is accentuated even further under effectiveness and timeliness criteria (b), owing to its perceived capacity for pre-emptive or responsive action in facing a sudden "climate emergency". Where safety is prioritised alongside effectiveness a somewhat different picture emerges (c), with air capture and storage performing the highest overall. Under affordability and timeliness criteria, stratospheric aerosol injection returns to a high performance, but alongside the somewhat more benign afforestation (d). Afforestation retains the coveted position of highest overall performance under affordability and safety, and timeliness and safety criteria (e, f).

These permutations demonstrate how different instrumental framings can serve to "close down" on certain geoengineering proposals (Stirling, 2008; Bellamy et al., 2012). Such closure leads to prescriptive policy recommendations that promote further research and investment in proposals that seem preferable given the narrow framings upon which they are built. This poses the risk of premature "lock-in" (Arthur, 1989) to particular future pathways that could instigate conflict and controversy between divergent values and interests. Geoengineering assessments should instead seek to "open up" option and policy choice by adopting broader and more diverse framings in order to guard against this lock-in (Stirling, 2008). Rather than providing prescriptive policy recommendations, the assessments would provide conditional recommendations that expose the framing conditions under which options perform.

The primarily narrow and technical problem definitions and criteria deployed in geoengineering assessments have recently prompted the use of a different kind of expert-analytic method; one that seeks to open up those, and other, inputs. Using a multi-criteria mapping methodology (Stirling and Mayer, 2001) and in defining the "problem" as one of responding to climate change in its

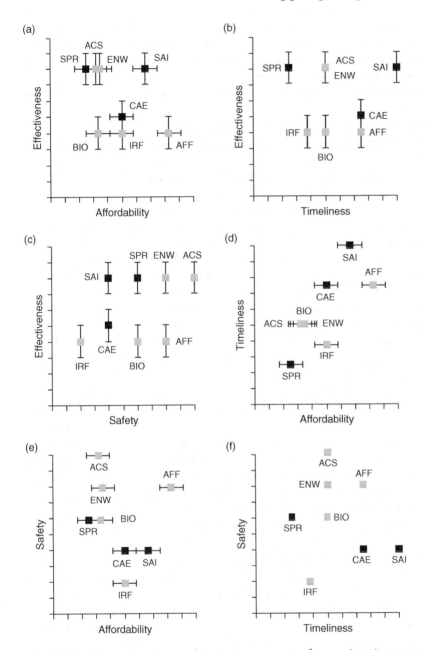

FIGURE 8.1 Permutations in a multi-criteria assessment of geoengineering proposals.

broadest sense, rather than of narrowly responding to insufficient mitigation or a climate emergency, a diversity of alternative options and criteria has been yielded (Bellamy et al., 2013). Both the range and depth of criteria have been diversified, under which a radically different picture of option performance emerges. The range of criteria groups has been opened up beyond "technical" issues to include those of "social" ones spanning politics, society, ethics, and co-benefits to show that geoengineering proposals, and stratospheric aerosol injection more acutely, are outperformed by mitigation alternatives. Moreover, the depth of criteria groups has been opened up within those "social" issues as well as the more "technical" issues of efficacy, environment, feasibility, and economics, to show the same pattern.

In seeking to open up the assessment of geoengineering further still, a small but significant literature in public participation has begun to emerge. The outcomes from participatory assessments (such as surveys or focus groups), however, are also susceptible to instrumental framing. The "climate emergency" framing has proven particularly potent in such settings, underpinning apparent public support for research into solar geoengineering in a cross-cultural survey (Mercer et al., 2011), and likely improving the perceived public acceptability of solar geoengineering in the UK-based Experiment Earth? (2010) public dialogue. A "naturalness" framing was also likely to have improved the perceived acceptability of particular geoengineering proposals in that same public dialogue (NERC, 2010; Corner et al., 2011). Framing air capture and storage as "artificial trees" and stratospheric aerosol injection as "no different to a volcano", for example, might constitute valid technical descriptions, but can also frame particular ends. Nevertheless, the naturalness framing has also emerged unprompted from publics themselves as an important element in forming their perceptions of geoengineering (Corner et al., 2013).

Broadening out and opening up geoengineering assessment reveals the complexities and uncertainties that are often reduced and hidden in narrowly framed assessments. For policymakers, this might make uneasy reading. Opening up assessment in these ways may appear to make decision-making harder rather than easier. However, this need not be the case. Neither expert-analytic nor participatory assessment methods are complete without the other, and so methods have been designed that integrate both elements. Deliberative mapping is one such method that we have recently developed and used (Bellamy et al., 2016), in combination with multi-criteria mapping (Bellamy et al., 2013), to evaluate geoengineering proposals alongside other options for tackling climate change. While the method seeks to map divergence of perspectives, it can equally map consistencies. Indeed, a remarkable level of consistency has been found across expert, stakeholder, and public perspectives, with geoengineering proposals being outperformed by mitigation alternatives.

We can draw a number of lessons for future research and policy from attempts at framing geoengineering assessment. First, the contextual isolation into which geoengineering has been placed must be overcome, by replacing narrow problem definitions, such as being faced with a "climate emergency", with broader ones such as "responding to climate change". This introduces alternative options spanning mitigation and adaptation that cannot be ignored. Second, the use of narrow, technical criteria must be overcome by expanding the range and depth of "social" as well as "technical" criteria. This second lesson must be supported by a third, in which the prevailing use of expert-analytic methods of assessment must be supplemented by, or integrated with (carefully framed) participatory methods that include a diversity of stakeholders and publics. This diversity will help hedge against the risk of "lock-in" to particular proposals that may at first appear preferable under certain framings. Taken together, these lessons argue that the framing of geoengineering assessments should be broadened out and opened up (Stirling, 2008; Bellamy et al., 2012). Indeed, as this perspective has discussed, some initial efforts have already been made in acting on these lessons, but much more remains to be done.

First published online as an opinion article in 2013.

9

PUBLIC PERCEPTIONS OF GEOENGINEERING

Rose Cairns

Given that geoengineering technologies remain for the most part hypothetical ideas, geoengineering research has been described as being at an "upstream" moment (Corner et al., 2012). This implies that, in contrast to more mature technologies which may already have become "locked-in" or resistant to change (Cairns, 2014a), the ultimate forms – if any – that these technologies might take in the future is still amenable to being shaped by the concerns and values of society. Thus there is a fair degree of consensus that eliciting public perceptions about geoengineering approaches is important (Shepherd et al., 2009; House of Commons, 2010; Rayner et al., 2013), and that it should happen while research in this area is at an early stage (Carr et al., 2013). However, eliciting, understanding, and representing what this group called "the public" might think or feel about geoengineering is not necessarily straightforward.

First, there is a variety of (stated and implicit) rationales for eliciting public views about potentially controversial technological developments like geoengineering. Fiorino (1990) distinguished between normative rationales (eliciting public perspectives about the possible development of technologies that would affect them is the right thing to do); substantive rationales (one gets substantively better, more socially robust decisions if one involves the public); or instrumental rationales (one should do it because it helps to achieve a given end – e.g. to get the public "on side" with regard to a particular socio-technical development). Also helpful to consider here is Stirling's (2008) distinction between the role of public engagement in closing down or opening up policy processes: is the aim of public engagement to reach a consensus or a majority view on geoengineering, or to justify a given policy commitment to a particular development trajectory? Or is it to open up the arguments, framings, and values inherent in these discussions to the widest possible range of perspectives, and illustrate the ways in which particular societal courses of action depend on the particular values, perspectives, or framings that are privileged? The view one takes on the purpose of eliciting public perceptions will affect not only where and how one looks for these views, but the seriousness with which different ideas or concerns are treated, and the degree to which ideas such as "representativeness" or "legitimacy" of particular views are felt to be of primary concern.

Second, different ways in which this imagined group called "the public" – a term which Laclau (2006) has called an "empty signifier" – are understood or constructed, have implications for the ways in which it would make sense to elicit their views. For example, some social scientists have questioned whether this thing called "the public" should be understood primarily as simply large numbers of private individuals, or whether it also makes sense to consider the many forms of "collective self-realisation of publics", including for example the views of social movements and civil society groups. Welsh and Wynne (2013) refer to these groups as "early risers" sensitive to normative social and cultural commitments, embedded, but often undeclared, in techno-scientific developments, which they regard as emergent public issues, stakes, and meanings. Similarly, some have argued that public discourses on a topic such as geoengineering contain a series of "latent meanings" likely to indicate the fault lines along which public opinion will likely divide in the future (Mooney, 2010). Some have even made a case for the representation of discourses rather than individuals in deliberative processes when the deliberative participation of all affected by a collective decision (such as is the case with decisions about geoengineering) is not feasible (Dryzek and Niemeyer, 2008). Following from these ideas, a number of authors have taken a discursive approach to examining, and subjecting to critical scrutiny, the range of discourses and framings of geoengineering in the public sphere, such as those expressed in academic publications or through a range of media channels (Nerlich and Jaspal, 2012; Sikka, 2012; Markussen et al., 2013; Porter and Hulme, 2013; Cairns and Stirling, 2014), including more marginal public discourses such as the belief in chemtrails, which, I argue elsewhere, may be revealing of concerns and values which are likely to resonate with other publics (Cairns, 2014b).

Third, given that geoengineering technologies do not yet (and may never) exist, attempts to elicit public perceptions more directly (e.g. through surveys, interviews, focus groups, or other deliberative exercises), need to overcome a number of methodological complications and potential pitfalls. Given that various studies concur that the awareness of the idea of geoengineering is still low among the majority of people (Mercer et al., 2011; Pidgeon et al., 2012), the way in which the topic is first introduced to people in order to elicit their opinions is crucially important. For example, a recent study illustrates the well-recognised impacts of framings on elicitation of public perceptions, showing experimentally the ways in which the use of natural metaphors to describe possible geoengineering approaches resulted in more positive perceptions, among individuals in the study, of those technologies (Corner and Pidgeon, 2014). Framing discussions of geoengineering in terms of a response to a climate emergency may also have a powerful impact on subsequent attitudes towards these approaches, putting participants in a disempowered position should they wish to express dissent or disagreement with the idea of pursuing geoengineering research (Parkhill and Pidgeon, 2011).

A number of attempts have been made to elicit public perceptions of geoengineering directly, including some large-scale surveys (Mercer et al., 2011; Pidgeon et al., 2012), and several more deliberative workshops, which have explored public perceptions towards geoengineering in general (NERC, 2010), towards stratospheric aerosol injection in particular (Macnaghten and Szerszynski, 2013), and towards a specific geoengineering research project, the SPICE project (Parkhill and Pidgeon, 2011). These different studies have produced interesting but sometimes divergent results based on their assumptions and methodologies. For example, based on the results of their large-scale survey, Mercer et al. concluded that there was what they considered a "surprisingly high" level of support for solar radiation management among the public, and classified 29 per cent of their sample (of 2893 people in the US, Canada, and the UK) as "supporters" of SRM and 20 per cent as "detractors" (Mercer et al., 2011). On the other hand, a smaller-scale deliberative study (based on seven focus groups involving between – six and eight participants in the UK) concluded that groups of supporters or detractors could not be so easily distinguished, and found that the process of deliberation in a group resulted in participants becoming increasingly more sceptical about SRM (Macnaghten and Szerszynski, 2013). Similarly other deliberative exercises have not sought to classify publics into supporters or detractors but highlighted the range of public concerns about proposed geoengineering technologies. Findings from these studies have echoed those of public engagement exercises around other novel or emergent technologies (Stilgoe, 2007), by illustrating the fact that public concerns often encompass but go well beyond narrow questions of feasibility, safety, or risk. For example, publics have raised questions like "who would control the technology?" or "what else might it be used for?", or "who would be accountable if things go wrong?" (Parkhill and Pidgeon, 2011).

Finally, it is worth highlighting that although the range of both discursive and more direct elicitation approaches to exploring public perceptions of geoengineering have produced some interesting findings and raised some important concerns and questions about geoengineering, one evident limitation of existing work is the narrow geographical diversity of the publics' views that have been explored to date. Although there have been some limited attempts to expand the conversation about geoengineering into different geographical contexts – for example the African, Chinese, and Indian workshops held by the Solar Radiation Management Governance Initiative (2013); or workshops held in Beijing and Delhi by the Climate Geoengineering Governance Project (for example, see CEEW, 2014), it is widely recognised that debates about geoengineering are overwhelmingly taking place in countries in the global North. This is clearly problematic, not least because the impacts of both climate change and proposed geoengineering interventions would likely disproportionately affect countries in the global South.

In conclusion, given the potential impacts of geoengineering interventions, opening-up discussions and debates about geoengineering research to as broad as possible a range of public concerns and perspectives is of crucial importance. But eliciting and representing the views or perceptions of "the public" towards geoengineering is a high-stakes activity, and it pays to be cautious in interpreting results, and alert to the politics of these processes. It is important to recognise the assumptions underlying the ways in which "the public" are constructed in any given study, and to be alert to different (stated and unstated) rationales for eliciting public views. In order to avoid the accusation that public engagement exercises are simply tick-box exercises aimed at legitimising existing research trajectories, the concerns of different publics need to be taken seriously. Although there are still minority voices within the geoengineering research community which depict public engagement in geoengineering research as a bureaucratic intrusion into the scientific process, or an attempt to shackle scientific freedoms, many geoengineering researchers demonstrate a high degree of reflexivity about their research and its potential implication (for example, see Stilgoe et al., 2013b). These researchers are very aware of the importance of understanding and engaging with the perceptions and concerns of different publics, and do not wish to carry out research without a social licence to do so.

First published online as an opinion article in 2015.

10

GEOENGINEERING AND THE INEXACT SCIENCE OF COMMUNICATION

Jeff Tollefson

On October 15, 2012, *The Guardian* broke a story about a "geoengineering scheme" conducted off of Haida Gwaii on the west coast of Canada. Led by the American entrepreneur and quintessential carbon cowboy Russ George, the Haida Nations village of Old Massett had dumped around 100 tonnes of iron into the ocean in hopes of spurring a phytoplankton bloom that would boost ocean productivity and benefit the salmon population that natives depend on for food. The venture was supposed to pay for itself via the sale of carbon credits that could be generated after some of those plankton – those that were not destined for the food web – died and sank to the bottom of the ocean (Lukacs, 2012).

Although the leaders of Old Massett were portrayed as victims, not villains, they were hardly pleased with the coverage. The village council had no planetary designs in mind at all, as suggested by the moniker it gave to the publicly funded entity that conducted the work: Haida Salmon Restoration Corporation. Regardless of their questionable scientific justification, carbon credits were merely a way to help pay the bills. But the storyline unfolded exactly as one might expect in newspapers around the world, focusing on the spectre of rogue geoengineering (Press Review: Iron Fertilization in Canada/Haida Gwaii, 2013). Only *Scientific American* bucked the trend with a lengthy interview of George himself (Biello, 2012), a businessman who has long pushed commercial ocean fertilisation. George repeatedly defended both the scientific theory and the implementation, but by this time few cared to listen.

Another scandal might have unfolded several months earlier when the Stratospheric Particle Injection for Climate Engineering (SPICE) experiment was cancelled in the United Kingdom. In this case, there were outstanding questions about the process for public engagement, but also a conflict of interest: a subset of the project's researchers had applied for patents covering the basic technology to be deployed in the device that would have sprayed water into the sky, apparently unbeknown to the funders. SPICE's leadership defused the situation by revealing the conflict and cancelling the field test (Watson, 2012), and most publications simply printed the news – naturally focusing on the conflict of interest – in a matter-of-fact manner (Cressey, 2012; Hale, 2012; Brown, 2012; Marshall, 2012). Had the story leaked into the press earlier, however, things might have turned out very differently.

Both stories provide a lens into the larger media machine. Journalists are often accused of seeking controversy and then pumping out shallow, sensationalist stories when they find it. This is often a fair assessment, but there are legitimate pressures and constraints that push all journalists in this general direction. My goal here is to explore some of these pressures and constraints, and how they necessarily affect coverage of a complex and compelling subject such as geoengineering.

Journalists look for a story

The first thing to keep in mind is that journalists look not just for news, but for news *stories*. Motivations vary among publications and individual reporters as well as their editors, but the desire to write stories that people might actually want to read is universal. Both cases cited above contain juicy revelations, but they differ in their dramatic appeal.

Haida Gwaii was the perfect story: would-be scientist dupes multiple governments and contravenes international protocol while seeking profits from unproven and perhaps dangerous science. Journalists, myself included, had little trouble finding scientists who were alarmed at the way the so-called experiment was conducted and sceptical of any results that might come of it. Virtually everybody distanced themselves from the project, including the Canadian government, which launched an investigation (CBC News, 2013), and the United States' National Oceanic and Atmospheric Administration, which said it was unaware of the nature of the work when it loaned George a small fleet of ocean gliders (Tollefson, 2012).

The revelations at SPICE had similar potential. Imagine the storyline had an intrepid reporter broken the story in the face of an intransigent and defensive scientific team: greedy geoengineers secretly seek to profit from planetary instability. It fits nicely into one of the themes that scientists and journalists have long feared. "Solar radiation management" might seem dauntingly complex, and understanding the consequences of deployment is indeed a supreme challenge. But the technology itself is not all that difficult, and any government – or even a wealthy individual with a deep wallet – could do it poorly or carelessly. No one has accused the SPICE scientists of any such thing, of course, but the storyline is already there.

As it happened, the SPICE leadership pre-emptively shone a light on the problem and announced the cancellation of the field trial (the basic research continued). Newsworthy, yes, but not exactly dramatic. "Geoengineering experiment cancelled due to perceived conflict of interest" read the headline in *The Guardian*

(Hale, 2012). My own publication, *Nature*, went with "Cancelled project spurs debate about geoengineering patents", using the news peg as an opportunity to explore some of the fundamental questions surrounding intellectual property (Cressey, 2012). Scandal was averted. Scientists coming clean and slapping themselves on the wrists, while newsworthy, simply doesn't rank as high as a rogue geoengineer implementing his master plan.

This is both natural and inevitable in a field that is long on theory, short on action, and imbued with end-of-times intrigue. Geoengineering mixes moral ambiguity and comfortingly conventional technology to fabulous effect. It is the steampunk of modern science and yet another Frankenstein for the 21st century – only this time the mad scientists are driven to counter technological indulgences that have pushed Earth systems to the brink of an unknowable regime change. We see this morality tale taken to one of its logical conclusions in the 2013 blockbuster movie *Snowpiercer*, which offers a portrait of humanity stuck on the technology train in a frozen world after a coalition of countries proceeds with stratospheric injections, which quickly prove all too effective. The experiment at Haida Gwaii fell well short of this scenario, but it pushed all of the right buttons.

Often scientists, not science, make the "news"

The problem facing both journalists and scientists seeking to educate the general public is that the science itself is hard to assess in mainstream publications. Numerous studies have advanced our understanding of geoengineering's potential as well as its potential knock-on effects, but conclusions are hard to come by. Various media outlets have picked up on recent results underscoring the danger that injecting aerosols into the stratosphere could alter weather patterns and ultimately the precipitation that people depend on. BBC summed it up this way in November 2014: "Climate fixes 'could harm billions'" (Shukman, 2014). Science doesn't always cooperate, but that's the kind of headline journalists love, often at the expense of the far more nuanced understanding that scientists are advancing.

When it comes to reporting science, "news" is also a relative term. The notion that climate change is difficult to get on the front page precisely because of its slow creep is not new, but this problem is only compounded when it comes to geoengineering. The BBC story above quoted Bristol University's Matt Watson, head of the SPICE project, calling geoengineering's effects – positive and negative – "really, really complicated". And then the bottom line: "We don't like the idea but we're more convinced than ever that we have to research it", Watson said (Shukman, 2014). Neither of these statements presents new scientific findings, but they do provide a bottom-line interpretation that helps orient the readers.

Let us consider these concluding passages to another article:

> Potentially catastrophic side effects make geoengineering schemes a dangerous gamble, asserts Robert Watts, an engineering professor at Tulane University and a global warming expert. "We don't understand the climate

system right now. It's a huge thing and it has enormous momentum," Watts said. "If you get the climate system going one way and all of a sudden you say, 'Whoops, I didn't mean for that to happen,' it isn't easy to turn it around." Still, even scientists who are very skeptical of geoengineering believe research should continue.

"It's important, in case the effects of global warming are more severe that we can deal with it in different ways," said Michael MacCracken, a geophysicist at Lawrence Livermore National Laboratory, near Oakland, Calif. "The melting of Greenland or the west Antarctic ice sheets would be the equivalent of adding 15 to 20 feet of sea level, so it starts to become difficult to figure out any natural way to reverse it." Said Watts: "Fifty or 100 years from now, if we have a crisis, these are all things you'd like to have in your hip pocket".

(Fagin, 1992)

Newsday published these words in 1992, following up on the National Academy of Science's landmark report (1991) a year earlier, dubbed "Policy Implications of Global Warming". Earlier references to space mirrors aside, the article could have been written today. Nearly a quarter of a century has passed, and while the science and even the political discussion of geoengineering have progressed, the basic "news" storyline hasn't changed.

From complexity and nuance, to headlines and column inches

As journalists, we are left to look for those moments where the landscape seems to shift in some significant way: a United Nations body weighing in on the policy landscape, a government decision to invest, or a major research initiative releasing its results. One such moment came in March 2010, when a small group of scientists and policy experts gathered to lay out the rules of the road for geo-engineering research at the Asilomar resort near Monterey, California. Given its storied place in the scientific discussion over recombinant DNA, the drama came naturally and the expectations were high.

What I encountered during five days of discussion at Asilomar was a prolonged and generally thoughtful debate among a variety of folks who ranged from engaged and wary to sceptical, fearful, and outright opposed. Climatologists debated with social and political scientists as well as humanitarians and senior bureaucrats on not just the practical and political considerations but also the ethical and moral implications of geoengineering. There were the usual discussions about how to differentiate among the different types of geoengineering, as well as well-intentioned-but-ultimately-doomed debates about whether "geoengineering" might be rebranded with a less sinister name (this was, after all, the "International Conference on Climate Intervention Technologies"). And then there were the deeper subjects: what would large-scale geoengineering signify for humanity? Given that we have already engineered our climate unintentionally, what is so scary about active and intentional climate management? Is it just the fear of unintended consequences? Or is it the act itself that strikes fear in our hearts?

It was an interesting affair, but the end result of such events, measured in terms of journalistic output, is often a bit of a disappointment. The immediate product of the meeting at Asilomar was a short document that endorsed further research into geoengineering while highlighting the need for engagement by the public and governments (MacCracken et al., 2010a); the final product, some basic principles for geoengineering research, would come later (MacCracken et al., 2010b). Given that the message emanating from Asilomar echoed the findings of previous geoengineering deep dives, most notably that of the Royal Society a year earlier, the meeting was more notable as a venue that brought people from different disciplines and with differing viewpoints together.

The multidisciplinary engagement was interesting, and at *Nature* we tried to capture some of the drama and the debate in 900 words, under the headline "Geoengineers get the fear" (Tollefson, 2010). The fact that it didn't necessarily advance the ball was not necessarily a bad thing. It would have been a surprise if it had. But it was an honest attempt to make sense of a controversial subject, and we treated it as such. And the conclusions echoed those in the stories above.

Yet scientists often complain that the nuances and often the results themselves get lost in a haze of oversimplification or outright sensationalism if they are covered at all. Both claims are often true, but even those stories that get it right necessarily omit far more than they include. This is to be expected: journalists are not court reporters. We have limited space to explain technical issues that easily fly over the heads of non-experts. Often, we are left to pick off an important study here or there that presents a curious result and use that as a peg to provide an update on the larger issue. This would seem to be the day-to-day modus operandi in many newsrooms.

My own coverage of geoengineering dates back more than a decade to a science column I wrote poking fun at the idea of deploying mirrors in space to block the sun's rays. I still find the subject intriguing, and I look for opportunities to dig deeper into the advancing science and related policy machinations. The goal is always the same: locate and explain those nuggets that will help average readers understand larger concepts and debates. It's a negotiation between new details and the larger context, between current drama (where we can find it) and boilerplate backdrop. When I go to my editor, I must be able to answer these questions: what is new? And, why do our readers care? Savvy experts understand this, and they curate their message accordingly.

Another way?

Within this context, it is hard to imagine the coverage of the experiment at Haida Gwaii going any other way. Everything was new, and our readers care because this is one small version of a nightmare future that has been explicitly laid out in discussions like the one that took place at Asilomar. It is possible to imagine a different ending to this story, but doing so requires imagining a different beginning as well.

Here we must go back to the original intent. The village council of Old Massett merely wanted to test the theory that salmon stocks can be increased by promoting occasional phytoplankton blooms. Some scientists had argued that the ocean dusting caused by natural volcanoes several years earlier helped boost ocean productivity, leading to a record salmon run (Buck, 2018). Could ocean fertilisation help accomplish the same thing? And might an experiment along those lines also provide valuable data for those seeking to understand the carbon cycling through such blooms? And if the science showed that some of the carbon was indeed sequestered at the bottom of the ocean, could carbon credits help pay for the venture?

There are some legitimate scientific questions in there. And the basic storyline of a village desperately seeking to preserve its way of life in the face of a changing environment, likely while struggling against environmentalists and international protocol intent on preventing further deterioration of the same, would provide plenty of drama. One can imagine ways where full and open engagement by a larger community of scientists could have produced a real experiment with results that the broader research community could have used. This would of course require an entirely different framework for public engagement and scientific outreach. It would have been difficult. But there is no doubt that it would have made for a good story.

11

"THIS IS GOD'S STUFF WE'RE MESSING WITH"

Geoengineering as a religious issue

Wylie Carr

"It's very dangerous to try and play God. This is God's stuff we're messing with. Historically speaki ng, any time we try to play God, we lose every time. That's what I think about it". This is how Peyton[1] – an Alaska Native, subsistence hunter, and wildlife specialist in Barrow, Alaska—described his impressions of geoengineering after hearing about it for the first time.

The "playing God" metaphor Peyton utilises is not novel; it appears time and again in discussions about emergent scientific and technological issues (Peters, 2007; Silver, 2006). In recent years, the phrase has been employed by journalists, philosophers, bloggers, and members of the public alike in reference to geoengineering (Carr et al., 2012; Farrell, 2012; Wagner and Weitzman, 2012). Ethicist Clive Hamilton suggests "playing God" resonates with both theists and atheists because it captures a sense of "humans crossing a boundary to a domain of control or causation that is beyond their rightful place" (Hamilton, 2013a, pp. 177–178). In other words, the metaphor indicates that the prospect of intentionally modifying the global climate evokes deeply held beliefs about the proper place and role of humans in the order of the cosmos. For many people these deeply held beliefs are religious in nature. Consequently, religious beliefs could play a critical role in future discussions about geoengineering at local, national, and international levels. This essay draws upon past social science research and recent interviews with religious individuals to argue that religion will play a role in public support for, or opposition to, geoengineering research in many countries.

In the United States, a large majority of the population claims a religious tradition, and religious groups exert significant social and political influence (Pew Research Center, 2012). Social science research indicates that religion also affects perceptions of emergent science and technology. Studies have found that religion negatively impacts belief in climate change and support for scientific research on issues like biotechnology and nanotechnology (Gaskell et al., 2005;

Maibach et al., 2009; Jelen and Lockett, 2014). Brossard et al. (2009) for example, found that in the US, religion played a more important role in shaping perceptions of nanotechnology than factual knowledge about the topic, with more religious individuals significantly less likely to support funding for research. In a similar study that compared the US with Europe, Scheufele et al. (2009) found that more religious countries like Italy and Ireland were similar to the US, exhibiting lower levels of public support for nanotechnology research than less religious countries such as Denmark and Germany. These studies indicate that religion provides a key perceptual filter or framework through which people interpret science communication and form opinions about emergent scientific and technological issues.

Religious beliefs will likely play a similar role with regard to geoengineering. As Kahan et al. (2015, p. 194) argue, "cultural values are cognitively prior to facts in public risk conflicts ... groups of individuals will credit and dismiss evidence of risk in patterns that reflect and reinforce their distinctive understandings of how society should be organized". In other words, people interpret information about science and technology within a particular cultural context and in light of the values that they hold individually and share with others. As a result, people accept or dismiss scientific evidence not on content, but whether or not it is framed in a way that aligns with or threatens their values and beliefs. Interestingly, Kahan et al. (2015, p. 19) found that the prospect of geoengineering better aligned with the values of certain cultural groups who view "human technological ingenuity as the principal means by which our species has succeeded in overcoming environmental constraints on its flourishing". Geoengineering could therefore make cultural groups who extol technological ingenuity more open to discussing climate change solutions. The exact opposite may be the case for religious groups.

Drawing on theological reflection as opposed to social science research, Forrest Clingerman (2012, 2014) suggests that religious responses to geoengineering are likely to emerge along a continuum that emphasises human fallibility at one end and human capability on the other. The fallibility perspective or narrative stresses the finitude of human knowledge and past examples of human hubris in trying to interfere with divinely ordered processes. As such, this perspective is likely to caution against geoengineering and suggest that attempts to modify the climate are likely to result in calamity. The capability perspective, on the other hand, views human nature and ability more optimistically. It suggests that despite past failures, humans should still employ their ingenuity and new technologies to address contemporary problems.

This tension between human capability and fallibility arose time and again in interviews I conducted with more than 100 individuals (like Peyton quoted above) in the Solomon Islands, Kenya, and Alaska (United States) about their views on geoengineering.[2] The majority of interviewees openly wrestled with whether or not they thought humans should try to modify the global climate. Roughly a quarter of the individuals I spoke with invoked religious beliefs or spirituality explicitly (and unprompted) as they grappled with this question. For example, Rachael,[3] an Alaska Native who directed an organisation dedicated to including indigenous people and knowledge in science, described her initial response to geoengineering in the following way:

> The indigenous side of me has an automatic reaction to messing with the creator's plans. I would venture to say that that's probably going to be the reaction from most indigenous communities. ... Our first concern is always the care of Mother Earth. That's the way we're taught from the day we can breathe, that we are the original stewards of the universe and our homelands. Our environment around us is meant for us to protect. ... I'm pretty certain if I had my elders sitting here, that they'd probably feel pretty much the same way.

Rachel's wariness of geoengineering is not necessarily surprising, or unique to religious persons. As David Keith et al. (2010, p. 427) have noted, "It is a healthy sign that a common first response to geoengineering is revulsion. It suggests that we have learned something from past instances of over-eager technological optimism and subsequent failures". A key difference for religious individuals is that this initial revulsion stems from deeply held beliefs about the proper order of the world, not just from anxieties based on past technological failures. In other words, in addition to concerns about human fallibility, religion may prompt a sense that humans simply should not intentionally interfere with the "domain of the Gods" (Donner, 2007).

This concern was, in fact, a common theme across all interviews where individuals discussed religion, regardless of location or religious tradition.[4] More specifically, all the individuals who referred to religious beliefs said their faith was a key reason they were dubious of geoengineering. This should not be taken as an indication that all religious traditions or religious individuals will react the same way. There will not be any single religious response to geoengineering. As Clingerman (2012, p. 208) points out, "Different religious traditions have different authorities, rituals, scriptures, historical contexts, and theological commitments. These result in a dialog between different approaches to technology, politics, and environmental concerns ...".

My point therefore is not to argue that religious individuals will interpret geoengineering in any particular way. Rather, the interviews I conducted suggest that the prospect of geoengineering evokes religious beliefs for religious individuals. Geoengineering confronts our perceptions of the proper place of humans in the world. For many, these perceptions are explicitly informed by religious beliefs about where humans stand in relation to nature, creation, and the divinities. While religion was not the focus of the interviews, and this sample is certainly not representative of the broader population in these three countries, these findings align with past research, and indicate that religion is a powerful frame that many people will draw upon to make sense of geoengineering.

So why has religion been largely absent from previous assessments of public perceptions of geoengineering? The most likely reason is that most studies to date have taken place in the United Kingdom (Corner et al., 2012), a "less religious country" (Scheufele et al., 2009; Pew Research Center, 2012). While 83 per cent of Americans claim that religion is either very important (60 per cent) or fairly important (23 per cent) in their lives, only 47 per cent of Britons say the same thing (with only 17 per cent indicating that religion is very important) (Ray, 2003).[5] However, there is ample evidence across existing research that geoengineering does raise questions about the concept of nature and how humans relate to it (Carr et al., 2012; Corner et al., 2012; Corner et al., 2013). In one UK based public engagement exercise in particular, Corner et al. (2013) found "messing with nature" to be a dominant narrative. They concluded that, "The wide variety of ways in which people ... conceptualised and debated the relationship between geoengineering and the natural world suggests that this will be a key factor determining public views on the topic as awareness of it grows" (Corner et al., 2013, p. 946). They also noted that it would be interesting to see whether or not religious individuals employed similar narratives.

Considering that nearly 85 per cent of the world's population claims membership of a religious group (Pew Research Group, 2012), religion will undoubtedly have much to say about geoengineering (Clingerman, 2012). Future research needs to explore the connections between religious beliefs and geoengineering more explicitly. Religion will not only affect individuals' perceptions, but also the tenor of public discussion, media frames, and even policy proposals in countries with large religious populations. As Corner et al. (2013) argue with regard to public perceptions of geoengineering more broadly, researchers and policymakers need to resist the temptation to dismiss religiously informed perspectives as irrational or anti-science. While certain religious traditions in countries like the US have a long history of clashes with science, concerns about geoengineering based on religious beliefs are fundamentally concerns about how humans should understand and relate to the world around them.

As a result, effectively engaging religiously informed perspectives means discussing deep-seated values – including the values that scientists and policymakers bring to the table. Geoengineering research could face stiff opposition if religious values are not taken into consideration. Furthermore, certain geoengineering approaches may be deemed wholly unacceptable from certain religious perspectives regardless. Advocates for geoengineering research should therefore bear in mind that for many people around the world, this is God's stuff we're messing with.

First published online as an opinion article in 2014.

Notes

1 Pseudonym.
2 Unfortunately there is not room here for a thorough discussion of the methods or results of this research. However, in brief, these three study sites were chosen to explore whether or not individuals in regions experiencing different impacts from

12

THE ETHICAL DIMENSIONS OF GEOENGINEERING

Solar radiation management through sulphate particle injection

Nancy Tuana

Introduction

The slow progress on an adequate policy solution in light of growing scientific understanding of the impacts of a warming world, limited success on efforts to mitigate the causes of anthropogenic climate change, and awareness of the high costs and the limits to adaptation have led some scientists and some policymakers to consider geoengineering as a potentially viable option to avoid "threshold responses" and dangerous climate change (Crutzen, 2006; Hansen et al., 2006; Wigley, 2006; Lenton et al., 2008; Fox and Chapman, 2011). Some approaches to climate engineering, indeed, even proposals to field test climate engineering technologies such as ocean iron fertilisation or increasing the reflectivity of the atmosphere to reduce the amount of sunlight that is absorbed, raise serious and complex ethical issues.

Given this, proposals to deploy geoengineering technologies, or even to field test some of them, must be accompanied by serious consideration of the ethical dimensions of geoengineering (Preston, 2012; Scott, 2012). However, adequate ethical analyses must be grounded in and arise from a robust understanding of the relevant scientific accounts of such technologies and their potential impacts, and an appreciation of any correlated uncertainties. Indeed, ethical analysis may require and point to needed scientific research in cases where there are coupled ethical-epistemic issues, that is, where ethical judgements require additional knowledge. Hence, the science and ethics of geoengineering are intertwined.

A comprehensive evaluation would have to address the ethical issues specific to each type of technology and approaches to implementation. For illustration purposes, in this chapter, I begin from concepts and dimensions of "justice" as a lens through which to filter ethical questions and to focus the analysis on a particular geoengineering technique: solar radiation management through sulphate particle injection (SPI). I base this choice on current assessments that:

i.   The technology needed to deploy this approach might potentially be scaled up over a relatively short period of time (a decade or less) given that we may be able to model it on existing technologies;

ii. Many believe it has the best potential of all geoengineering approaches for cooling the planet rapidly and inexpensively;

iii. Given a) and b), it is being given serious consideration by a significant number of scientists and policymakers;

iv. SPI, unlike most geoengineering approaches that focus on absorbing carbon from the atmosphere (carbon dioxide management technologies), has the potential to rapidly create a novel atmospheric state, with significant regional or global climatic effects, but one of which we have limited knowledge (Morrow et al., 2009, p. 2); and

v.   It would, on its own, not mitigate the high concentrations of greenhouse gases. For these reasons, SPI is arguably the approach most likely to be considered for deployment and partial deployment for testing but is also the geoengineering approach most fraught with ethical issues.

The domain of ethical and coupled ethical-epistemic issues regarding SPI is very large. Indeed, an entire chapter could be written on each one of its many aspects. My aim in this chapter will be rather to map an agenda of research questions by working to identify some of the many issues in need of additional analysis regarding the ethical issues surrounding SPI. While I do not claim that these questions are exhaustive, the number and complexity of questions that emerge signal the importance of work on such topics prior to any decision as to whether to deploy SPI, as well as decisions about field testing.

Projected SPI impacts on temperature, precipitation, and society

Understanding the relevance of dimensions of justice requires an appreciation of the complexity of the impacts of SPI, and thus must be informed by an appreciation of our current scientific accounts and, in turn, work to identify value decisions embedded in the scientific analyses. Furthermore, ethical analysis must also be informed by an adequate analysis of the uncertainties in the science, and, ideally, partner with science to identify uncertainties due to missing domains of information that are required for adequate ethical analyses, as well as clarify which domains are due to deep uncertainties that are impossible to resolve through research. To frame the justice dimensions of SPI, this section provides an admittedly brief overview of some aspects of our current scientific understanding that are relevant to justice dimensions of SPI, while acknowledging that these details have been covered in greater strength elsewhere and since this chapter was written.

Modelling studies suggest that SPI, over the course of an 80-year simulation, likely could (if it were chosen as the target objective) stabilise average global-mean surface air temperatures at levels approximately plus or minus half a degree from the temperature at which the SPI activities were initiated (Ricke et al., 2010). Depending on how such an SPI intervention is done, it is possible that it could result in regional temperature disparities (Govindasamy et al., 2003; Lunt et al., 2008; Brovkin et al., 2009; Kravitz et al., 2014). Moreover, if SPI is not matched with serious mitigation efforts over time, the temperature differences between regions may increase such that within six to seven decades "there is often no scenario that can place a region back within one standard deviation of both its baseline temperature and precipitation" (Ricke et al., 2010, p. 538). In other words, in a non-mitigation scenario, different regions will likely experience different "climates" the longer the forcings continue, and the forcing scenario needed to return one region to the designated baseline (e.g., a late 20th-century climate) may be different than that required for another region the longer SPI continues.

Some models also indicate potential impacts of SPI on the hydrological cycle (Bala et al., 2008; Lunt et al., 2008; Robock et al., 2009; Irvine et al., 2010; Curry et al., 2014). The changes in precipitation may not be consistent across regions, and some regions may experience greater deviations than others in amount and seasonality of precipitation than others depending on the intensity and duration of the SPI (Irvine et al., 2010; Ricke et al., 2010). Given the potential for distinct regional differences in the response to different levels of SPI, the decision as to the choice of the optimal target for SPI (temperature or precipitation) may prove regionally dependent.

Precipitation changes due to SPI (as with those due to greenhouse gas driven climate change) could also have disproportionately negative impacts on those regions already experiencing high levels of poverty that have historically not significantly contributed to climate change. Based on natural experiments such as the Mount Pinatubo eruption and information from computer simulations, some models suggest that SPI could decrease average annual precipitation in Africa, South America, and south-eastern Asia (Matthews and Caldeira, 2007). Such changes in regional precipitation could compromise basic rights of individuals in these regions by resulting in food and water insecurity (Brewer, 2007; Robock et al., 2008).

Regional differences in precipitation impacts, however, are not the only component of ethical issues raised by SPI targets. Irvine et al. (2012) found there to be important trade-offs between SPI targets that would aim to reduce the rate of temperature change vs. the rate of sea-level rise (see also MacMartin et al., 2013). They demonstrate that addressing sea-level rise would require significantly greater forcing than would be required to stop surface warming. The greater forcings required for targeting sea-level rise, however, carry a significantly higher risk of abrupt or disruptive cooling (ibid.).

The temperature and precipitation impacts of SPI will not only impact human well-being, but will also affect ecosystem and species well-being (Naik et al., 2003).

Furthermore, since CO_2 levels may continue to increase during SPI, ocean acidification could continue to be a serious problem (Raven et al., 2005; Fabry et al., 2008; Doney et al., 2009). Ozone depletion may be another side effect of SPI (Rasch et al., 2008; Tilmes et al., 2008; Moan et al., 2008; Pitari et al., 2014), which would increase the risk of human health impacts, as well as the well-being of various species and ecosystems.

While a comprehensive review of current and rapidly evolving scientific understanding of SPI and its potential is far beyond this chapter, the diversity of issues identified in this brief introduction already presents numerous ethical questions.

The question of justice and differential impacts

There are at least five dimensions of justice relevant to SPI, including distributive justice, intergenerational justice, corrective justice, ecological justice, and procedural justice. While it is valuable to understand and examine each of these justice vectors, they almost always intersect in the case of SPI, greatly complicating the ethical analysis. In this section, I will discuss each aspect of justice relevant to SPI separately in order to clarify the types of issues relevant to each domain, following each description with a number of coupled ethical-scientific research questions in order to clarify the nature and range of coupled ethical-scientific issues that must be addressed to determine whether or not SPI could be considered as an ethically responsible choice. It is my intent to clarify the salience and complexity of issues of justice that are relevant to SPI, catalyse appreciation of the complexity of the intertwined ethical-scientific issues, and urge that this work is incorporated into the SPI research agenda. In addition, it is imperative that we recognise that these dimensions of justice often intersect and, indeed, at times conflict.

Distributive justice involves the principle that harms and benefits of an action should be fairly or equitably distributed. Strict egalitarian theorists argue, for example, that distributive justice requires that harms and benefits be shared equally. Rawlsian inspired difference theorists allow that differential impacts are justified on the condition that the least well off are better off than they were previously (Rawls, 1971). In the case of SPI, the appropriate measure would be spatial – namely, do all regions of the Earth equitably benefit and are any of the resulting harms fairly distributed?

Research questions:

1. Are the temperature and/or precipitation disparities caused to some regions by SPI outweighed by the benefits to other regions? Is it possible for all regions to benefit somewhat from small amounts of SPI? What levels of confidence of impacts would be required to make this judgement?
   a. What would be a morally salient difference between regions that might justify a positive response? For example density of population;

uniqueness of species; vulnerability of region to temperature and pre-cipitation change.

2. How should existing climate conditions be considered when assessing the impacts of SPI-induced changes in climate? For example, should a dry region getting drier be treated in the same way as a wet region getting drier? Can we actually disaggregate the SPI impacts from other impacts on such conditions (e.g. land use changes, etc.)?

3. If temperature and precipitation impacts resulting from SPI cannot or should not simply be aggregated, what is the best way to quantify them? For example:

   a. In weighing impacts, what is the ethically responsible way to com-pare positive/negative impacts of temperature changes with those of precipitation?

   b. How are regional variations regarding the risks of higher temperatures vs. modified precipitation to be weighted?

4. Are there certain harms from SPI that could not be justified regardless of the benefits to others? For example:

   a. Compromising basic human rights, such as food, shelter, and health, of populations in some regions?

   b. Loss of a culturally significant way of life?

   c. Loss of citizenship (climate refugees) or in extreme cases an entire country's sovereignty (an entire country becoming uninhabitable due to extreme weather conditions such as flooding)?

   d. What levels of confidence in the probability of such impacts would be required to decide against using SPI?

5. Given that temperature and precipitation differences between regions are very likely to increase if strong mitigation is not pursued in parallel with SPI deployment, could there be a time limit to SPI after which continually increased SPI would no longer be ethically justifiable?

6. Is the deployment of SPI to avoid "threshold" responses more just than the deployment of SPI to prevent global mean temperatures from rising above a certain degree? And if so, why?

7. Are there situations in which we would have a moral responsibility to use SPI rather than allow humans, other species, and ecosystems to suffer the harms of unremediated climate change? What would constitute the condi-tions that would make SPI a moral imperative? Are we able to measure such "thresholds"? What level of confidence in SPI's effectiveness regarding such a target should be required?

8. How do we weigh social benefits against risks to individuals? What meas-urements are actually possible and at what levels of confidence?

Intergenerational justice takes into consideration the impacts of SPI on future generations. Many see this version of justice as similar to that of distributive jus-tice but adding a temporal measure by comparing harms and benefits to current populations to those of future generations. However, even when attention is given to the impacts of a geoengineering approach to anthropogenic global warming upon future generations, an adequate ethical analysis must combine attention to impacts on future generations with attention to the spatial dimension of the impacts, as would an account of distributive justice, in those instances when future impacts might disproportionately benefit or harm different regions.

Research questions:

1. How would the harms to future generations of long-term SPI deployment compare to other scenarios (such as business as usual, mitigation efforts without SPI, etc.)? Are we able to effectively model such scenarios? In such models, what is included as a benefit and what is included as a harm? Under what scenarios do the comparative benefit/harm ethically justify SPI deployment?
2. Are there SPI scenarios (intensity/duration) that are ethically untenable regardless of the benefit to current generations because they would put future generations at ethically unacceptable levels of risk? What level of confidence would be required to make this judgement?

Corrective justice diverges from the egalitarian approach of typical accounts of distributive justice, by arguing against an "aggregate" measure of benefits and harms and embracing a desert-based measure, which holds that harms and benefits ought to be shared among persons according to the degree they deserve those harms and benefits. Through this lens, whether or not the impacts of an action are, just requires that we consider the extent to which individuals or groups are morally deserving of those impacts. In the case of anthropocentric global warming, desert-based accounts of justice are often based on responsibilities for emissions. A key element is historical contributions to climate change that are disproportionately due to the activities of a group of industrialised nations which benefited from the industrialisation and land use changes that led to high emission levels but at the cost of the well-being of other countries. Indeed, this form of corrective justice is recognised in the framework of the UNFCCC's "polluter pays" principle.

However, historical responsibility is not the only relevant measure. The inequities in emissions, and thus their inequitable contributions to the problem of anthropogenic climate change, continue to be an issue. For example, based on information from 2008, the United States emitted approximately four times the amount of CO_2 as India and more than 90 times the emission of CO_2 of Bangladesh (UN, 2008). If responsibility is correlated with population, however, issues of desert get more complicated. For example, while China's total CO_2 emissions were approximately 20 per cent higher than those of the US in 2008, average individual emissions in China are significantly lower than individual emissions in the US, where on average each citizen emits almost three times more CO_2 in comparison to individuals in China. But using average emissions also ignores that individual emission levels are linked to economic class standing, with the poor even in the highest emitting countries often having low emissions and the wealthy in the lowest emitting countries often living lifestyles that result in GHG emissions similar to wealthy individuals in high emitting countries (Harris, 2010). Corrective justice can also be decoupled from responsibility for the causes of greenhouse gas emissions, and focus, as does prioritarianism, on the position that justice requires that benefits to the worst off should be given more weight than benefits to the better off (Parfit, 1997).

And finally, just as intergenerational justice requires attention to the spatial distribution of harms and benefits at different times in the future (i.e., which regions at a particular time are likely to benefit or be harmed by the action), corrective justice must also embrace a temporal dimension, considering not only desert and culpability in the present case, but also projecting into the future to adequately apply such an account of justice.

Research questions:

1. In choosing targets for SPI (temperature/sea-level rise, etc.) should the decision be based on the greatest overall positive impacts or should those regions most negatively impacted by climate change be those regions that benefit most from SPI? Can we actually target in ways that would allow us to meet what is determined to be just targets? And how should historical and/or contemporary responsibility for greenhouse gas emissions be factored into the decision concerning which targets are the most just?

2. Should a "polluter pays" principle be applied to any responsibility for compensation for harms of SPI so that those most responsible for anthropocentric climate change become the most responsible? And if so, which of the following measures are ethically relevant? Should these measures include discounting?
   a. Historical responsibility for greenhouse gas emissions
   b. Per capita emissions
   c. A country's total emissions.

3. Temperature and/or precipitation changes will impact regions differently dependent on the general resilience/vulnerability of that region. How should the political-economic situation of a region be factored into the analysis of the impacts of SPI? Would such targets be technically feasible?
   a. Does justice require that benefits to poorer regions of SPI deployment provide these regions with greater benefits and fewer harms?
   b. Should those countries historically responsible for any political-economic vulnerabilities of a region be responsible for compensating vulnerable regions negatively impacted by SPI?

4. Are those regions that benefit most from SPI then responsible for compensating those regions which benefit less or which are negatively impacted? And if so, how should historical and/or contemporary responsibility for greenhouse gas emissions be factored into the decision about compensation?

5. Should current generations compensate future generations for the impacts of SPI?

6. What are ethically acceptable forms of compensation? For example, if a nation loses sovereignty because its land has been made uninhabitable for SPI, does compensatory justice require providing its people with comparable land where they can claim sovereignty?
7. Should regions likely to be harmed be provided financial support for adaptation prior to or during SPI deployment? And who is responsible for providing that support? For example:
   a. Historical responsibility for greenhouse gas emissions
   b. Benefits from SPI.

Ecological justice is a non-anthropocentric dimension of justice, which includes consideration of the impacts on non-human life and on ecosystem sustainability. Here, the emphasis of ethical analysis is the harms and benefits of SPI upon animals, plants, and ecosystems in general. As with the impacts of geoengineering on humans, the effects on other life forms and on ecosystems are dependent on the intensity and the length of SPI. And as with humans, animals, plants, and ecosystems in some regions will likely benefit, while others will likely be harmed. For example, geoengineering would not address the problem of ocean acidification. We know that high levels of CO_2 alter ocean chemistry and can negatively affect the shell formation ability of marine calcifying organisms such as corals, with subsequent impacts on the ecosystem level (Doney et al., 2009). SPI will have the effect of lowering ultraviolet radiation levels, which might enhance plant health, but as this effect is likely intertwined with other changes, both to precipitation and to seasonal climate, the benefit might not be to the plants currently growing in a particular region, but rather to new species that may or may not be beneficial to ecosystem health.

Research questions:

i. How do we weigh the moral standing of non-human species and/or ecosystems in comparison to that of humans in order to determine how to balance ethical responsibilities to current and future human populations with ethical responsibilities to current and future species and to ecosystems from SPI impacts?
ii. Are there certain harms to species or ecosystems from SPI that could not be justified regardless of the benefits to humans? What levels of confidence would be required to make this judgement?

Procedural justice focuses once again on the human domain, but in this case on how to ensure that decision procedures are ethical. Following Rawls, many have argued that in order to be procedurally just, all those affected by the decision must have the ability to contribute to the decision process or have their interests represented (Rawls, 1971; Müller, 1999; Grasso, 2007). Others argue that procedural justice also requires that the rationales for the policy decisions be transparent and public, the decision process be based on relevant ethical principles, and the process allow for a mechanism for appeal and regulation to ensure fairness (Daniels and Sabin, 1997). In the instance of SPI, procedural justice issues are relevant in a number of domains, including who makes the decision about whether to test or implement SPI, when to stop testing or deploying SPI, as well as what should be the target of SPI.

Research questions:

1. Since SPI will very likely affect all nations, must any just decision process for implementation be an international process?
   a. Is there an existing body like the United Nations that would be appropriate for this process? Does it provide sufficient representation of all those likely to be impacted?
2. Is the nation state the ethically relevant representative group for making a just decision about SPI? If not, what would be?
3. How widespread must agreement be on SPI deployment for it to satisfy the demands of procedural justice?
4. What principles and procedures are best suited for making an ethical decision about SPI targets?
5. If there are individuals, groups, or nations who do not consent to SPI deployment, are they thereby more deserving of compensation for resulting harms?
6. Is there ever a condition in which it would be ethically acceptable for one group (e.g., a nation) or a small federation to make the decision to geoengineer without consultation with other groups/nations?

The applicability of the various dimensions of justice in the case of SPI arises from the well-recognised fact that SPI deployment could have serious side effects. "A world cooled by managing sunlight will not be the same as one cooled by lowering emissions" (Keith et al. 2010). But it is also linked to the fact that the speed and intensity of SPI deployment options correlate both to different climate "remediation" impacts as well as to different distributions of harms and benefits (Irvine et al., 2012).

The question of intentionality

One of the reasons SPI is seen as raising serious ethical issues is the question of intentionality. As noted by many ethicists writing on climate change, climate change is often not viewed as an ethical issue because it does not embody the characteristics of a paradigm moral problem. According to Jamieson, "a paradigm moral problem is one in which an individual acting intentionally harms another individual; both the individuals and the harm are identifiable; and the individuals and the harm are closely related in time and space" (Jamieson, 2007, p. 1). While Jamieson argues persuasively that climate change nonetheless raises ethical issues, the link between moral responsibility and harm for SPI is arguably clearer and stronger due to the fact that those acting will be acting with knowledge that their actions have a high probability of violating the basic rights of people in some regions and could potentially be damaging to the rights of future generations as well as to non-humans.

Jamieson refers to various types of geoengineering, particularly large-scale projects like SPI, as "intentional climate change". While we now know that many human activities from agricultural practices to energy choices are impacting the climate, the fact is that SPI has as its primary intention to modify the climate, and it would be done knowing that there are various risks and highly probable harms, as noted above. Jamieson argues that this places large-scale geoengineering projects like SPI in a different ethical domain in that the decision to modify the climate would be the intent of the actions, and thus the consequences of the action to deploy, including the unintended harms, would have a stronger ethical tie to the action.

Research questions:

1. Is intentionally creating novel climates for the purposes of alleviating at least some of the harms of anthropocentric climate change ethically more problematic than business as usual greenhouse gas emissions now that we know that these emissions contribute to anthropocentric climate change?
2. SPI will only, at best, lessen some of the negative impacts of anthropocentric climate change and not the causes, thereby allowing greenhouse gases to continue to accumulate. Under what conditions, if any, is it ethically permissible to deploy SPI alone, that is, without mitigation efforts?
3. Does the fact that SPI intentionally creates novel and unpredictable climates entail that the individuals or groups who elect deployment are ethically responsible for any resulting harms?
4. If the intentionality of SPI entails greater ethical responsibility for resulting harms, does this result in a greater responsibility for those who agree to deployment to compensate those who are harmed?

The question of risk and uncertainty

Intentionality often raises what has been called the "principle of double effect". According to this principle, an action is ethically acceptable even if those acting cause or allow something bad as long as a) no evil is intended as an end or a means; and b) the potential harm is not out of proportion with the anticipated good. However, the relationship between intention (to slow down the aggregate warming) and consequences (the harms and benefits of SPI) is made significantly more complex due to the fact that there are uncertainties linked to the probabilities of various impacts of SPI. Sidgwick, for example, influentially argues that "it is best to include under the term 'intention' all the consequences of an act that are foreseen as certain or probable; since it will be admitted that we cannot evade responsibility for any foreseen consequence of our acts by plea that we felt no desire for them" (Sidgwick, 1907, p. 202).

To include all the consequences of SPI deployment "that are foreseen as certain or probable" puts us in the domains of risk management and decision-making under uncertainty. At least some of the uncertainties relevant to SPI can be mitigated through additional scientific research. However, the question of testing itself raises a series of complex ethical questions. I will reserve a discussion of these concerns for the next section and focus here on some of the ethical dimensions of decision-making and risk management under conditions of uncertainty.

Various principles have been advocated by those working on the ethical dimensions of risk management (Wikman-Svahn, 2012). One is the principle of justification, which requires that for any action that entails the risk of harm, the benefits should outweigh the harm. While necessary, theorists argue that additional principles are required for an action to be ethically justifiable. One is the principle of optimisation which implies that the likelihood of harm, the number of people exposed, and the magnitude of the harms "should all be kept as low as reasonably achievable, taking into account economic and societal factors—meaning that the level of safety should be the best under the prevailing circumstances maximizing the margin of benefit over harm" (González, 2011, p. 2). A third principle is that of individual protection, namely that the risk incurred by any individual should be restricted. This principle is designed to go beyond calculations of aggregate harms and focus attention on the magnitude of harms to individuals, in order to determine if there are limits that must be imposed on risk to individuals, for example, risk to satisfaction of basic needs.

Each of these principles requires transparency regarding the relevant value judgements that would be involved in its application. For example, the principle of justification would require weighing harms and benefits, which will likely be significantly different in kind. It will also have to take into consideration the various ethical dimensions noted above of differences in harms/benefits to various regions and between current and future generations. While perhaps providing guidelines for ethical decision-making, principles such as these still leave unsettled large domains of ethical analysis.

In addition to value judgements such as these that are involved in decisions concerning acceptable levels of risk, what to count as a harm and to whom/what, or how to weigh different types of harms and/or benefits, the question of uncertainty raises additional ethical concerns. Various types of uncertainties are relevant to SPI. There is significant epistemic uncertainty in that we currently have incomplete knowledge about the impacts of SPI. Research on the impacts of SPI has been to date limited, and much of the research that has been done has been on natural experiments or modelling. To gain additional knowledge would likely require additional research, including at least partial deployment for testing, that itself raises coupled epistemic-ethical concerns as will be discussed below. Epistemic uncertainties can be reduced with sufficient time and resources, but ethical issues are relevant to how long we can wait to resolve such uncertainties before deciding whether or not to act. There is also ontological uncertainty, or what some have called deep uncertainty, in that aspects of the interactions between SPI and the natural systems are complex and non-linear and thus unpredictable. These are uncertainties that are inherent in the complexity and coupled nature of the problem, and will not be mitigated with additional research.[1] And third, SPI involves ethical uncertainty, in that there are different values and principles concerning how to weigh the harms/benefits of SPI, different judgements about the seriousness of those harms/benefits, different interpretations of who and what are to be included in the domain of moral standing (e.g. are non-human species and ecosystems to be included), and uncertainty about what future generations would view as the most salient ethical values or principles.[2]

The ethical dimensions of decision-making under conditions of uncertainty is a new domain of ethical analysis (Caplan, 1986; Tannert et al., 2007), but one essential to SPI. Robust decision aking (Groves and Lempert, 2007; Lempert and Collins, 2007; Bryant and Lempert, 2010; Hall et al., 2012) and dynamic adaptive policy pathways (Haasnoot et al., 2013) are two relatively recent approaches to decision-making under conditions of uncertainty, including ontological uncertainty. In both instances, the authors of these approaches appreciate the need to have ethical analyses closely intertwined in the analyses. While these approaches are not specific to SPI, they offer strategies for identifying ethically responsible ways to make decisions under conditions of uncertainty, including all three of the domains of uncertainty noted above. While uncertainty clearly makes responsible decision-making more difficult, it is a condition underlying many of our most trenchant global issues and thus works to identify ethically responsible decision-making approaches.

Research questions:

1. To what extent must we reduce epistemological uncertainty in order to make an ethically responsible decision about SPI deployment?
2. Does ontological uncertainty about SPI deployment entail following a precautionary principle and not deploying at all or unless the potential harm of not doing so would clearly outweigh the uncertainty of doing so?
3. What are the best ways to manage ethical uncertainty about SPI?

The question of testing

As noted above, there is significant epistemic uncertainty surrounding SPI deployment, which some believe can and should be lessened through scientific testing. Indeed, it has been argued that there is an urgent need for research into geoengineering options such as SPI and that this research should go beyond modelling or the analysis of natural events, such as volcanic eruptions, and include field studies (Keith et al., 2010; Dykema et al., 2014). This type of testing is believed by some to be a necessary and ethically responsible step prior to partial and/or full deployment of SPI for geoengineering and is seen as providing the basis for evaluating SPI technologies, testing the response of the system, and exploring possible unintended consequences. However, this position assumes that testing can occur that is a) significantly different than deployment and b) can lessen the epistemic uncertainties concerning SPI impacts.

Tuana et al. (2012) argue that testing in the way defined above may not be possible for a variety of reasons. First, there are major uncertainties in climate models such as vertical mixing in the ocean (Wunsch and Ferrari, 2004; Goes et al., 2011), evolution of polar ice including ice sheets and glaciers (Meehl et al., 2007), radiative feedbacks in the atmosphere (Bony and Dufresne, 2005), and clouds and precipitation which would be highly sensitive to SPI deployment. Second, non-linear feedbacks in the climate system can result in bifurcations of the system leading to abrupt shifts or transitions between states, such as the shutdown of the ocean's meridional overturning circulation, resulting in markedly different climate conditions. This is relevant to the question of testing as posed in that there may be a significant difference between small forcings of the kind that would be deployed for testing and the forcing levels and time trajectories needed for intentional climate modification. As forcing increases, the climate system could reach a threshold where it transitions to unstable conditions. In such a case, the SPI is happening in a significantly different climate state than that in which it was tested. Third, the system may exhibit hysteresis, or strong memory, in which reducing the forcing after the testing may not return the system to the original climate. Fourth, there will likely be delayed system responses to forcing. We know that different time scales govern ocean and atmosphere circulations, such that oceanic responses to SPI forcings may not manifest for years to decades longer than atmospheric responses. Because of this, impacts from SPI deployments for the purpose of testing may not be fully realised by the climate system until long after stopping the testing. Given the above noted variables, the type of learning projected from small-scale deployment for testing may not be possible. Given also that these experiments can have negative impacts, both the ethical and the scientific justification for conducting such experiments are at issue.

With these concerns there is a variety of ethical concerns regarding field testing of SPI. Here I will identify some of the ethical issues directly related to field testing per se, but it is important to underscore that many of the ethical issues noted about regarding issues of distributive justice, intergenerational justice, compensatory justice, ecological justice, and procedural justice apply to field testing as well given the "side effects". Note that these are examples of coupled scientific-ethical issues.

Research questions:

1. What can be inferred from the limited-scale experiments about the potential of a full-scale experiment, and what cannot?
   a. Is it possible to estimate the large-scale system response from a small-scale field test?
2. Will this knowledge be adequate for making an ethically responsible decision? Will this knowledge be sufficient to warrant the risks of field testing?

3. What "side effects" will result from field testing and can they be predicted?
4. What scientific and ethical knowledge is required to responsibly decide whether to start SPI field testing?
   a. What is the basis for deciding on acceptable risk levels for field testing?
   b. What measures of impacts would be used to determine that the costs of field testing are higher than the benefits of field testing and should be halted?
   c. What level of learning would justify risks of side effects?
5. What is the boundary between field testing and deployment?

Political risks of SPI research and testing

Another cluster of ethical issues concerns the psychological or political impact of SPI research or field testing. Some researchers have raised the concern that geoengineering research might pose a moral hazard by causing people to be less concerned than they otherwise would be with respect to the risks posed by climate change (Preston (ed.) 2012). Some have begun to question whether or to what extent SPI research would impede research into other responses to climate change or reduce the political will to mitigate greenhouse gas emissions (Bunzl, 2009). There are also concerns that conducting SPI research would lead to unregulated, unilateral, or self-interested uses (Victor, 2008). Others have argued, to the contrary, that a credible threat of unilateral SPI might strengthen global mitigation efforts to avoid potential costly side effects of SPI (Millard–Ball, 2012).

Conclusion: the centrality of ethics for SPI

Given the potentially harmful impacts of SPI there is widespread agreement that SPI deployment raises important ethical issues (Keith, 2000, pp. 277–278; Crutzen, 2006; Kiehl, 2006; MacCracken, 2006; Morgan and Ricke, 2010; Robock, 2008; Royal Society, 2009; Tuana et al., 2012). The 2009 Royal Society Report, to give just one example, affirmed that "it is clear that ethical considerations are central to decision-making in this field" (Royal Society, 2009, p. 39) and concluded that "the acceptability of geoengineering will be determined as much by social, legal and political factors, as by scientific and technical factors" (p. 50). However, in closing, it is important to stress that the ethical analyses of SPI are not simply an addition to the scientific analysis, to be put into play once the scientific research is complete. Ethically significant decisions are often embedded in the scientific analyses themselves, as well as in how scientific models represent impacts and vulnerabilities.

Ethical analysis is dependent upon and must be intertwined with robust and sound scientific knowledge and effective and ethically responsible decision-making tools. As we have seen from the above discussion, since SPI testing and deployment could involve non-trivial risks of harm across many dimensions such as time, space, species, and socio-economic status (Crutzen, 2006; Goes et al., 2011; Svoboda et al., 2011), an epistemically and ethically sound characterisation of the underlying probabilities and risks requires a well-integrated analysis spanning fields such as Earth sciences, statistics, and economics (Goes et al., 2011). Hence, many of the ethical issues identified in this essay require additional and targeted coupled scientific-ethical research to ensure that we are developing epistemically responsible knowledge about geoengineering and comparing it to mitigation options.

While acknowledging the importance of the Royal Society's recognition that ethical issues are central to decision-making, what is in fact required goes beyond ethical analyses of the science of geoengineering. It is essential that the ethical analysis be coupled with scientific analysis by including ethicists within scientific research teams in order to infuse ethical analyses into the science of geoengineering. This will, of course, require scientists and funding agencies alike to recognise the importance of such work and provide ample resources for coupled ethical-scientific analyses within SPI research. I close then with the admonition that this important field of study be strengthened prior to and included in considerations of the feasibility of SPI deployment as well as pre-deployment for testing.

First published online as a working paper in 2013.

Acknowledgements

This research was conducted with the support of National Science Foundation Grant Number 1135327 and richly informed by collaborations with Klaus Keller, Ryan L. Sriver, Peter J. Irvine, Jacob Haqq-Misra, Toby Svoboda, and Roman Olson. I would also like to gratefully acknowledge the reviewers whose insightful comments helped improve this article.

Notes

1 For greater detail on the nature of epistemological and ontological uncertainty, see Walker et al., 2003.

2 My taxonomy is somewhat similar to that of Tannert et al., 2007; however, my interpretation of ethical uncertainty diverges significantly from their account of subjective uncertainty, which includes what they call moral uncertainty, but define differently than what I mean by ethical uncertainty.

PART III
Geoengineering experiments
Early days

Introduction

While Part I demonstrated that geoengineering concepts are not entirely novel, the spate of scientific assessments conducted over the last decade have also demonstrated that there remains considerable uncertainty about their feasibility, effectiveness and potential side effects (Royal Society, 2009; NRC, 2015a, 2015b; EuTRACE, 2015). This section of the book explores the scientific research agenda surrounding geoengineering technologies, looking both forward towards potential experiments in the real world, and back at recent experiments that provide some insight into the challenges ahead.

Mark Lawrence, together with co-author Paul Crutzen, who launched the contemporary discussion of geoengineering as a potential response to climate change, open this section, describing in Chapter 13 how the scientific research agenda surrounding geoengineering evolved in the years shortly after Crutzen's 2006 article, and reflecting on where it may go in the future. In Chapter 14, Alan Robock and Ben Kravitz continue in a similar vein, examining the use of climate models and natural experiments (like volcanoes) to study the potential climatic effects of proposed geoengineering interventions. In both chapters, the authors reflect on the value and limitations of model studies, and the need for and risks of real-world field experiments to develop better scientific understanding of proposed geoengineering technologies.

The Earth systems modelling of SRM has expanded considerably since the writing of these contributions: in publication volume, in the types of technology examined, in the location, rate, term, and scale of deployment scenarios, in the granularity of regional and global repercussions, and in the understanding of model uncertainty (Irvine et al., 2016).[1] The political orientations of modelling have also increased. Some have noted that early and foundational simulations (in

GeoMIP, see footnote 1) were bluntly designed to induce large perturbations in the climate system and to learn more about Earth systems processes rather than reflect what might be politically desirable; hence, more moderately paced scenarios might form a better basis for policy deliberations (Keith and MacMartin, 2015). SRM modelling has also received increased attention as a "design problem" that highlights capacity (at least, in models) to shape impacts according to climate objectives by altering the parameters of deployment (MacMartin et al., 2013). These movements reflect ongoing tensions surrounding modelling as an imperfect basis for policy, and they remain unresolved (see Wiertz, 2015 for an interrogation of SRM modelling as an "inventive tool", where scientists envision and frame different possibilities for SRM deployment through the design of modelling scenarios).

A prominent tension still revolves around the threshold of risk between – and by extension the necessity of moving from – desk to fieldwork. Robock and Kravitz's chapter reminds us that researchers disagree as to whether this question can (or should) be answered by the natural and applied sciences. The next two chapters pick up this important thread by examining past attempts at geoengineering field experiments. In Chapter 15, Jack Doughty explores three small-scale field experiments related to solar radiation management (SRM) geo-engineering that cover a surprising range of intents and outcomes. In Chapter 16, Holly Buck looks in detail at one of the most infamous geoengineering field experiments to date: the ocean iron fertilisation (OIF) experiment conducted off the west coast of Canada in the summer of 2012. The insights from these various cases are myriad and complex, raising difficult questions of sovereignty, ownership, economics, and public engagement. Since then, there has been little movement – and perhaps little appetite – for more fieldwork, in the absence of a more permissive environment. One initiative that may move the needle (at least for SRM) might be a small-scale stratospheric chemistry test, led by prominent researchers based at Harvard, which is currently in development (Dykema et al., 2014). What is clear is that ensuring societal concerns are effectively taken into account in future geoengineering experimentation will require considerable attention and effort.

The section closes with Chapter 17 by Duncan McLaren who explores the research agenda for direct air capture (DAC) technologies that remove carbon dioxide directly from the atmosphere using chemical processing. Throughout the chapter, McLaren points out the delicate balance between constructive development of DAC to enable likely necessary negative carbon dioxide emissions in the future, and overly optimistic projections of DAC capabilities and low cost that dissuade timely action on emissions reduction.

Note

1 These movements rely strongly, though not exclusively, on the efforts of the Geoengineering Model Intercomparison Project mentioned in both chapters, for which a full list of publications can be found at their website (see Bibliography; see also Kravitz et al. 2015 for the framework of the latest phase of activity).

13

THE EARLY EVOLUTION OF CLIMATE ENGINEERING RESEARCH

Mark G. Lawrence and Paul J. Crutzen

The emerging climate engineering (CE) debate will be fed by scientific information – and likely by misinformation as well. This leaves us with the question: how has the scientific discourse around CE evolved? And how will it likely evolve in the coming years? Until recently, only a very limited number of scientists ventured to seriously investigate climate engineering, and this research has been predominately theoretical, mostly with computer simulations. No field tests of continental- or global-scale CE have been carried out to date, rather only small-scale tests of various aspects of CE, such as carbon dioxide (CO_2) uptake by phytoplankton, underground CO_2 sequestration, and brightening of marine clouds.

Here, we would like to present a scientist's perspective of the early evolution of CE research. In doing so, out of the various developments over the past few decades, we have chosen to focus on three major topics which have been instrumental in driving the recent evolution and setting the stage for further developments: 1) ocean (iron) fertilisation, 2) the set of articles published together in *Climatic Change* in 2006 on stratospheric sulphate injections, and 3) the advent of coordinated modelling activities to study climate engineering. These topics, especially the first one relative to the latter two, are a bit disconnected; this reflects the early activities in this field, which cover a very broad range of proposed CE techniques, and the field still currently has no form of international coordination or top-down organisation. Following this, we will also give an outlook to how climate engineering research might continue to evolve. Although these selected topics are mostly focused on natural science aspects, some of them touch on social, philosophical, and governance aspects, and we would like to emphasise up front that these interdisciplinary aspects have been and will remain crucial to the overall discourse.

Ocean (iron) fertilisation

The first intense, international studies of any type of CE began in the early 1990s with ocean iron fertilisation (OIF), a method associated with the carbon dioxide removal (CDR) branch of CE techniques.[1] There have now been large international research projects on the topic, which included over a dozen ocean iron fertilisation experiments between 1990 and 2009, numerous supporting studies with numerical models, and even a few companies that grew rapidly based on speculative sales of carbon credits on the voluntary market; a detailed overview of the scientific developments in the field of OIF over the past two decades is available in Williamson et al. (2012), and an overview of the activities of the commercial interests that have been involved in trying to develop or prepare the pathway for marketing OIF techniques is given in Eli Kintisch's 2010 book *Hack the Planet*.

The research on ocean iron fertilisation has provided a very mixed picture of the effectiveness of carbon drawdown and sequestration into the deep ocean. However, a few points have become generally clear. First, ocean fertilisation can apparently very effectively increase the downward flux of carbon under some conditions, and fail completely under other conditions; there are many complexities that are not well understood. Moreover, quantifying the actual amount of post-fertilisation deep-ocean sequestration is generally extremely challenging and highly uncertain, though possible under some limited oceanic conditions. Finally, the plankton blooms are very likely to be accompanied by substantial undesired side effects on the marine environment and the atmosphere.

The ocean fertilisation experiments and analyses led to discussions by scientists and stakeholders about not only the environmental advantages and disadvantages but also political and ethical concerns. Until recently, there was no legal framework for regulating ocean fertilisation. In the mid-2000s, efforts were begun to better connect the natural science analyses with other disciplines, especially law, as well as with the practitioner's perspective from the side of policy advice, in order to help better place the scientific results into a language that would be more supportive of the development of policy measures (Rayfuse et al., 2008). Along with pressure from NGOs, the first such interdisciplinary analyses helped lead to a resolution in 2008 by the London Convention and Protocol[2] prohibiting ocean fertilisation for commercial purposes, but allowing approved, basic science experiments, subject to a relatively restrictive assessment framework for environmental impacts. Among all the proposed CE techniques, ocean fertilisation was the first and is still the only one, which in principle has a basic regulatory framework for research and implementation. However, the effectiveness and enforceability of this framework is now being put to the test, following a covert implementation of iron fertilisation off the coast of Canada in the summer of 2012, for which legal action is currently being investigated (see Fountain, 2012 for a representative media account).

Articles published in *Climatic Change* in 2006

The research on OIF over the last two decades, both in terms of field experiments and computer model simulations, have demonstrated that there was no broad taboo in the scientific community against scientifically investigating ocean fertilisation. Quite the opposite was the case for the other form of CE, "solar radiation management" (SRM) (Shepherd et al., 2009). This taboo became very apparent in the discussions around an editorial essay about SRM using stratospheric sulphate aerosols published by Paul Crutzen in 2006, along with five commentaries published together in the same issue (Cicerone, 2006; Kiehl, 2006; Bengtsson, 2006; MacCracken, 2006; Lawrence, 2006). Extensive, harsh criticism was expressed by many members of the community, centred upon two main concerns. First, SRM schemes neither address CO_2 emissions nor other environmental impacts such as ocean acidification and might thus detract from these serious concerns. Second, there may be a "moral hazard", in that researching SRM may unintentionally help to legitimise it, raising hopes that there will be an alternative "quick fix" solution to climate change, and thereby derailing the efforts to reduce CO_2 emissions.

Attention to these concerns resulted in careful statements being included in Crutzen (2006), indicating the caution needed in considering climate engineering, including a discussion of the possible side effects, and the many scientific, legal, ethical, and societal issues, and finally noting that "the albedo enhancement scheme should only be deployed when there are proven net advantages and in particular when rapid climate warming is developing, paradoxically, in part due to improvements in worldwide air quality". This cautious encouragement of research was also reflected in nearly all of the five invited commentaries published alongside Crutzen (2006). The experiences with OIF noted in the previous section were very influential in leading to this encouragement as stated in Lawrence (2006), especially having recognised how a good scientific knowledge base was important in the support of and eventual formulation of a resolution on governing OIF. It is, however, unclear if that was an important motivation in the other calls for research, and it is also unclear whether this conclusion can really be applied to other types of CE (determining that would be an extensive research project in and of itself).

After the publication of Crutzen (2006) and the commentaries, the openness of the community towards research on SRM changed rapidly. Prior to 2006, there were only a very limited number of studies of SRM, and the core climate modelling community around the CMIP (IPCC) simulations had not yet addressed it in their models. In the first half-decade since then there have already been numerous publications looking at various aspects of climate engineering using models, with many scientists being motivated by the recognition that "geoengineering is being discussed intensely, at least outside of the formal scientific literature, and it is not going to go away by ignoring it or refusing to discuss it scientifically" (Lawrence, 2006). One of the most important highlights of lifting the taboo on SRM research is that there are now substantial efforts to coordinate community-wide simulations with multiple models.

Community-wide multi-model studies

The first publications on climate engineering were single model studies of various aspects of various CE techniques, for instance the global distribution of the effect of turning down the solar constant, or determining the amount of stratospheric sulphur injection that would be needed to produce a certain radiative forcing and what this would imply for changes in the ozone layer. These resulted in a rather mixed picture, with little clarity as to which results from which models to trust, or in many cases, how to even compare the results from different set-ups of studies. Nevertheless, despite the uncertainty in models, they are the tool that currently needs to be relied on to gain an initial impression of the potential effectiveness and side effects of various proposed CE techniques; field tests are limited in terms of the information they could give, since at small scale it is unclear how they would scale up given the vast heterogeneity in the global Earth system, and at a scale large enough to be confident in the scaling the field "tests" would have more of the character of a small-scale implementation.

A major step forward to help reduce the uncertainty in the analyses of the model simulations began around 2010. Two representative projects in this space were IMPLICC[3] (which ended in 2012) and GeoMIP (which is ongoing today).[4] These projects decided to use the same main scenarios for stratospheric sulphur injections and solar constant modifications (emulating mirrors in space) for their simulations. The IMPLICC and GeoMIP simulations (Kravitz et al., 2011) built on the simulations of CMIP5 (Coupled Model Intercomparison Project, 5th generation) for the IPCC Fifth Assessment Report. In that sense, these multi-model projects generally relied on the climate and Earth system models in the form that they are available and used for the CMIP5 simulations. IMPLICC and GeoMIP have contributed to understanding how to set up useful CE simulations that are comparable across a wide array of models, and to improving on the robustness in the interpretation of the model results, as well as the sense of uncertainty in these results. However, the projects are not contributing significantly to the development or improvement of the models themselves.

Outlook – how will climate engineering science evolve further?

While a major part of the future evolution of CE research will extend on some of the trajectories outlined above, especially the expansion of community-wide multi-model studies, a highly uncertain direction of development involves future field testing of CE. The field tests of ocean fertilisation were discussed above; there is now a great debate in the scientific community as well as between civil society organisations and various researchers regarding field testing of SRM. Without singling out members of the community here, it has become evident that there are several who contend that it is necessary to have field tests in order to get beyond the uncertainties in models, while others claim it is far too premature and may lead to a backlash, potentially even inhibiting future research of other climate-relevant processes like cloud microphysics, which could be mistakenly perceived as being done in relation to CE. Generally, governance of CE research distinct from other, non-CE basic research will face the challenge that these will often be difficult to distinguish from each other in terms of the forms in which they are carried out, and intention (i.e., whether for the purpose of developing an understanding of CE or not) is even more difficult to determine and distinguish.

Beyond the natural science research which has been the focus here, two extremely important further developments are underway: first, the effort to connect across the various disciplines involved in understanding CE, including the natural sciences, economics, psychology, philosophy, political sciences, and law; and second, the challenge of "trans-disciplinary" research, the co-generation of knowledge by researchers and stakeholders. Many CE studies have already been published in the separate disciplines, with only a few interdisciplinary studies so far, and even less trans-disciplinary work. However, substantial efforts are being made at building these bridges between the disciplines and between researchers and stakeholders, in particular through several large projects such as the European Transdisciplinary Assessment of Climate Engineering (EuTRACE) that have taken on this challenge in recent years.[5] This will be especially relevant in leading to sensible governance – which is also likely to include governance of the scientific research itself as the field continues to evolve.

The primary message we have hoped to bring out here is that research on climate engineering is evolving extremely rapidly, and while some developments can be foreseen, it is extremely difficult to know how this research will impact policy development, and especially how it can be targeted to effectively inform the social dialogue around CE, without risking derailing the basic efforts towards mitigation and adaptation measures. Because of the importance of carrying out research on CE in an effective and responsible manner, as well as a larger context of global efforts towards sustainable development, a deeper understanding of the evolution and future pathways of CE research is needed.

First published online as an opinion article in 2013.

Notes

1 OIF is the idea that enhancing growth of phytoplankton in the oceans (e.g., by fertilising with iron or other nutrients) would cause them to draw down additional CO_2 from the atmosphere for photosynthesis, locking up the carbon in biomass, some of which will sink to the deep oceans following death or ingestion and excretion.
2 The London C/P is the international maritime convention regulating the dumping of waste in the ocean.
3 "Implications and Risks of Engineering Solar Radiation to Limit Climate Change", http://implicc.zmaw.de/ IMPLICC completed its progamme of work in 2013, and a final summary report can be found on its website.
4 "Geoengineering Model Intercomparison Project", http://climate.envsci.rutgers. edu/GeoMIP/. GeoMIP has produced dozens of publications (http://climate.envsci.

14

USE OF MODELS, ANALOGUES, AND FIELD TESTS FOR GEOENGINEERING RESEARCH

Alan Robock and Ben Kravitz

Geoengineering[1] has been suggested as a theoretical response to anthropogenic global warming (Crutzen, 2006). However, geoengineering has not been conducted, so there are no data or observations of it. How then can geoengineering be studied? One obvious technique is to use global climate models ("indoor" research) to simulate various proposed geoengineering schemes, such as adding aerosols to the stratosphere to reflect incoming sunlight or adding sea salt to marine stratus clouds to brighten them. Since these two techniques mimic volcanic eruptions and ship tracks, another suggestion is to study those phenomena as analogues to geoengineering. There have also been several suggestions for field experiments, as well as some small-scale tests ("outdoor" research), to learn about geoengineering. In this article, we review these different research methods, commenting on their utility, safety, ethics, and governance. We also discuss natural analogues for geoengineering, such as the 1991 eruption of Mt. Pinatubo and the observation of ship tracks, highlighting both their utility in learning about the effects of geoengineering and their limits in providing knowledge. As we will demonstrate, geoengineering research is inseparable from climate research.

Climate models

Climate models are an obvious tool for geoengineering research. In these models, it is possible to perturb the climate system with various patterns of stratospheric aerosol injection or marine cloud brightening and investigate the climate system response. The vast majority of geoengineering research so far has been with climate models, and these investigations have proven to reveal much about the effects of certain methods of geoengineering, as well as the fundamental underpinnings of climate system response to perturbations. The Geoengineering

Model Intercomparison Project (GeoMIP) (Kravitz et al., 2011) has resulted in a wealth of studies that examine the climate response to several different, specified scenarios of stratospheric geoengineering.[2] The large voluntary participation of climate modelling groups from around the world in this project, and the opportunity to compare their responses to standardised forcing, clearly demonstrate the utility of this type of research. New climate modelling experiments, including the design of several new experiments for marine cloud brightening (Kravitz et al., 2013) and cirrus thinning (Kravitz et al., 2015), promise that much additional knowledge about geoengineering will be provided in the near future.

Natural and anthropogenic analogues

Volcanic eruptions are a clear natural analogue for stratospheric sulphate aerosol geoengineering (Crutzen, 2006; Wigley, 2006; Robock et al., 2013). Robock et al. (2013) discuss this issue in great detail, and here we only summarise some of the points. The observation that large volcanic eruptions cool the planet was one of the original motivations for suggesting geoengineering (Crutzen, 2006). For example, the eruption of Mount Pinatubo in 1991 cooled the planet by roughly $0.5°C$ (Soden et al., 2002) by injecting between 10 and 20 Mt SO_2 into the stratosphere. However, volcanic eruptions are an imperfect analogue for stratospheric geoengineering because of confounding effects of volcanic ash, because volcanic eruptions are into a clean stratosphere, and because of differences between continuous and impulsive injection of material into the stratosphere. The difference in the longevity of the injected particles means that climate system responses with long timescales, such as oceanic responses, would be different between volcanic eruptions and long-term geoengineering, but rapid responses, such as seasonal responses of monsoon circulations and precipitation would be quite similar (MacMynowski et al., 2011), and the volcanic analogue would be appropriate. Geoengineering in particular seasons could increase the effectiveness of geoengineering (MacMartin et al., 2013), decrease the amount of direct interference in the climate system through geoengineering, and make the analogue of volcanic eruptions more applicable.

Nevertheless, volcanic eruption analogues already reveal many things about the potential effects of continuous stratospheric sulphate aerosol clouds. Some examples include cooling the surface, reducing ice melt and sea-level rise, increasing the land carbon sink (Mercado et al., 2010), reduced summer monsoon precipitation (Oman et al., 2006; Trenberth and Dai, 2007; Robock et al., 2008), destruction of stratospheric ozone that allows more harmful UV at the surface (Tilmes et al., 2008; Pitari et al., 2014), whitening of the sky (Kravitz et al., 2012), reduction of solar power (Murphy, 2009), damage to aeroplanes flying in the stratosphere (Bernard and Rose, 1990), and impacts on remote sensing (Strong, 1984). Study of past and future large volcanic eruptions promises to help answer additional questions, including the growth and distribution of sulphate aerosols, impacts on ozone and on cirrus clouds, and the effects of increased water vapour (because of a warmer tropical tropopause) in the stratosphere.

Ship tracks, where there is a clear cloud signal resulting from the injection of aerosols from the ship exhaust (Christensen and Stephens, 2011), can indicate the effectiveness of increasing the brightness of marine boundary layer clouds through the injection of aerosols such as sea salt (Latham, 1990). Robust relationships among changes in precipitation, cloud albedo, and cloud coverage have not yet been established from observations, but both careful data analysis and greater observational capability may help to better understand how cloud brightness, lifetime, and extent would respond to particle injections. Aerosols in clouds can produce many more subtle effects in addition to the visible ship tracks, and cloud albedo is not always enhanced by increasing the aerosol concentration (Christensen and Stephens, 2012).

Field tests (outdoor experiments)

It is almost certainly true that some questions about geoengineering can only be answered through outdoor field tests. However, the claim that there is a need for these field tests (Parson and Keith, 2013) would need to be substantiated. Without clear goals for such research, demonstration that the research is safe, and externally evaluated, monitored, and regulated governance, such outdoor research is unethical (Robock, 2012). Some proposed research involves emission of pollutants, such as SO_2, into the atmosphere, and the emissions need to be regulated to prevent environmental damage.

Some of the aspects of geoengineering proposals could be tested outdoors at a small scale that would provide useful information while not significantly increasing risk to the environment. For example, can an aeroplane be constructed that can take a tank of SO_2 gas (or other sulphate aerosol precursor) into the lower stratosphere, spray it out, and create a cloud of sulphuric acid droplets of a desired size distribution? If so, how much would it cost and how dangerous would it be to the operators of the system? Rough estimates made so far suggest that such an apparatus would not be expensive (Robock et al., 2009), but field tests (e.g., Dykema et al., 2014) could calibrate these estimates. However, this experiment would not test whether such a cloud could be produced that would limit the growth of the aerosols (Heckendorn et al., 2009). Such an outdoor test would have to be done at a scale that would essentially be actual implementation of geoengineering (Robock et al., 2010).

The distinction between small- and large-scale tests can be somewhat blurry, but caution can still be used in determining whether a field test should be conducted. First, experiments should be designed to meet a clear goal and in such a way that minimises risks to other parts of the environment. Second, the benefits of conducting the experiment should outweigh the risks. Field studies in many branches of science, such as with weather modification, pesticides, or genetic crop modification, proceed with the knowledge that the experiments may cause harm, but the knowledge gained is deemed to be more beneficial than the potential risks. Such determinations and weightings of benefit and risk are made by external regulatory agencies, and the experiments are monitored closely. These governance structures would also be necessary for geoengineering field studies to be conducted ethically (Robock, 2012; Parson and Keith, 2013).

Outdoor experiments should not be conducted if there is another, less risky way of obtaining the same information. For example, volcanic eruptions, climate model simulations, and previous studies of the radiative effects of aerosols have shown that layers of aerosols can intercept solar radiation, so there is no need to conduct an additional field experiment (Izrael et al., 2009) to do the same thing. And some risky experiments should never be done, even if they promise to provide information to test theoretical results. The prime example of such a societal decision is that nuclear weapons are no longer tested, even underground, except by North Korea.

The relationship between geoengineering research and climate science

All the methods mentioned above for conducting geoengineering research were developed for general climate studies. The scientists conducting the research are climate scientists. Climate models continue to be developed in centres around the world as a representation of our best knowledge about how the climate system works. Field campaigns are routinely conducted to observe and measure the atmosphere and how it is changing. A network of satellites, air-based measurements, and ground-based measurements continually provides information about the current climate state. Any climate modelling studies of the effects of geoengineering will use climate models, and any measurement of the effects of field tests or deployment would use the current observation network.

Moreover, some of the fundamental questions about the effectiveness of geoengineering are intimately related to fundamental questions in climate science. How does the climate respond to changes in radiative flux? How do aerosols and clouds interact? What observation system is needed to determine the effects of the next large volcanic eruption? The study of the climate and the study of geoengineering are tightly linked. Conducting geoengineering research has proven to be very useful in understanding the fundamental processes that govern climate behaviour, and in turn, a better understanding of the climate will promote a better understanding of the effects of geoengineering.

Summary

Transparent research on geoengineering is an essential part of the discussion wherein the benefits and risks of geoengineering can be determined. There is little reason to regulate curiosity-driven indoor research, provided it does not cause any dangerous environmental effects. However, outdoor experiments should be assessed to determine whether they are dangerous, and they should be regulated, even if these experiments are for scientific purposes. There is precedent for governing dangerous human inventions, such as ozone-depleting substances and nuclear weapons. Such mechanisms are based on widely accepted norms of environmental protection and independent regulation. These structures are necessary to weigh the benefits of knowledge about geoengineering against the risk of not knowing.

First published online as an opinion article in 2013.

Acknowledgements

Alan Robock is supported by US National Science Foundation grants AGS-1157525, GEO-1240507, and AGS-1617844. Ben Kravitz is supported by the Fund for Innovative Climate and Energy Research (FICER). The Pacific Northwest National Laboratory is operated for the US Department of Energy by Battelle Memorial Institute under contract DE-AC05-76RL01830.

Notes

1 In this paper we only address solar radiation management and use the term "geoengineering" to specifically refer to those sets of technologies.
2 For an ongoing list of GeoMIP publications, see http://climate.envsci.rutgers.edu/GeoMIP/publications.html

15

PAST FORAYS INTO SRM FIELD RESEARCH AND IMPLICATIONS FOR FUTURE GOVERNANCE

Jack Doughty

Outdoor field research studies self-identifying as or closely related to solar radiation management (SRM) technology have already been carried out by scientists (Welch et al., 2012). Exploring these experiments illustrates the varying degrees to which scientists have wrestled with the ethical, social, and environmental governance concerns that have been raised by many (for example, see Corner and Pidgeon, 2010). Exploring how these past outdoor research projects were carried out (or not, as is the case with SPICE) highlights the need for existing governance to be adapted to ensure that these concerns are addressed. This paper will look specifically at three examples of outdoor research from recent years:

- Yuri Izrael's solar radiation experiment carried out in Russia with an uncertain degree of scientific credibility and an unclear assessment process
- E-PEACE, an experiment which while not identifying as SRM resulted in clear implications for marine cloud brightening (MCB) technology post hoc
- The SPICE project's cancelled balloon deployment experiment, whose "stage gate" process was interested in exploring a wide range of SRM governance issues.

As scientists develop proposals for future SRM experiments (Keith et al., 2014), these past projects provide key insights into the challenges that governance must navigate.

There are currently few governance frameworks which deal explicitly with SRM research. While international bodies for environmental protection exist, they tend to be geographically limited or based on particular issues and would need to undergo considerable adaptation to apply to SRM field research (Bodle, 2010). National and regional governance mechanisms such as environmental impact assessments have been suggested as a potential governance mechanism for

SRM experiments (ibid.); they are already required in the London Convention and Protocol for ocean fertilisation experiments (IMO, 2013). The experiments examined here have relatively negligible environmental impacts and involve no climatic response. Proposed experiments for the immediate future are likely to carry similarly small environmental risk (Dykema et al., 2014). The challenge for governance of any future experimentation is to address the wide range of social and political concerns that have been pinpointed as well as any environmental impacts. These case studies illustrate that not all scientists are engaged with these questions and that those who are face their own set of challenges.

The Yuri Izrael-led solar radiation experiment

Reported in *Mother Jones* as the first example of an outdoor SRM field research (Mooney, 2009), the experiment was conducted in August 2008, 300 miles south-east of Moscow and involved spraying sulphuric aerosol into the troposphere in order to study its effect on solar radiation. The experiment was led by Yuri Izrael from the Institute of Global Change in Moscow, with a team of Russian scientists. While there is little available information on what prior assessment procedure took place, there has been no report of the experiment encountering any major obstructions either at a local, national, or international level, or encountering any challenges from environmental non-governmental organisations (ENGOs) (Caviezel and Revermann, 2014, p. 197). The results of the experiment indicated, according to the team, that the scaled up technology has the potential to effectively counter global warming (Izrael et al., 2010). The experiment received comparatively little attention at the international level despite the results being published in a paper: "Field Experiment on Studying Solar Radiation Passing through Aerosol Layers". Though the experiment has been criticised as lacking in scientific credibility (Meleshko et al., 2010), the fact that it was carried out in the first place raises questions over how any future governance could prove effective operating within complex political spheres.

The researchers reportedly conducted model experiments before taking their research outdoors in order to examine whether there would be any negative environmental impact (Welch et al., 2012). They came to the conclusion that since the project was not likely to have any detrimental impacts on the environment, it would be fine to proceed (ibid.). There are few information sources available on the experiment. An article in *Wired* magazine describes the experiment as having been conducted over a two-square-mile area outside the city of Saratov on the Volga River – 300 miles south-east of Moscow – and that the helicopter and truck was provided by officials from the Russian Federation (Kintisch, 2010a). Sulphuric aerosols were injected from the car chassis and helicopter at heights of around 2.5km (troposphere rather than stratosphere), with two detectors on the ground measuring the solar radiation, wind velocity, temperature, humidity, and pressure (Izrael et al., 2010). The paper by Izrael et al. states that due to the weather conditions there was "a high degree of variability of the data obtained" (ibid.). Nevertheless, the paper ascertains that it is "principally possible to control solar radiation passing through artificially created aerosol formations in the atmosphere", marking this as the first outdoor fieldwork to promote SRM technology as a potentially effective option.

This experiment was carried out in a considerably different political climate to those examined in the North American and European contexts. Izrael, often described as having been a key scientific confidant of Prime Minister Putin[1] as well as a prominent member of the Russian Academy of Sciences, was the PI for this experiment. A renowned advocate of SRM technology and a sceptic of man-made climate change (MosNews, 2005), he previously had published an open letter to Putin calling for SRM as an appropriate step against global warming, arguing "in order to lower the temperature of the Earth by 1–2 degrees we need to pump about 600,000 tons of aerosol particles" (ibid.).

With Izrael having been such a strong supporter of SRM technology and the experiment having taken place in a relatively secretive state, others have cast doubt over the scientific value of the experiment (Meleshko et al., 2010). This lack of credibility seems to have lessened the impact and saliency of this experiment in the wider world. Nevertheless, it raises important issues for future SRM governance. An atmospheric chemist in Izrael's institute stated that they were hoping to conduct larger experiments, possibly by using aircraft over large areas, assuring that this "would be a very local experiment over Russia, only over Russia" (Kintisch, 2010a).

The difficulty any governance mechanism would have to address is how to effectively cover national level experiments operating under limited levels of transparency. While the experiment's SRM intentions were clearly stated, it underwent no international scrutiny process, with the research team reportedly not viewing any as necessary (Welch et al., 2012). With Russian attitudes portrayed as relatively pro-geoengineering, some are concerned further experimentation could take place (Lukacs et al., 2013). Debate over governance of SRM in Western countries is ongoing,[2] but currently international regulation can do little to stop secretive and geographically vast nations such as Russia carrying out atmospheric experiments. Even if a suitable international body was to set up restrictions there is no guarantee that countries that demonstrate scant regard for such treaties would uphold them. The most interesting factor about this experiment is that despite being the only experiment of the three that was fully carried out and which self-identified as SRM, it gained the least attention from opponents of experimentation. This raises important questions over how governance of research can operate when ethics and geographical locations can result in a lack of international transparency.

E-PEACE

The Eastern Pacific Emitted Aerosol Cloud Experiment (E-PEACE), conducted in August 2011, had no stated intention of examining SRM technology.

Despite this, it led to data results with clear implications for marine cloud brightening (MCB) (Russell, 2012). Examining ship tracks' interaction with clouds, the experiment used smoke generators to produce emissions comparable to ship tracks. Despite the team's focus on cloud-aerosol interaction, the results also found that ship tracks have a significant impact on radiative forcing. This implication has received criticism from some SRM researchers who argue that experiments "engaged with geoengineering more obliquely" do not pay heed to the social and ethical considerations that others have and are engaged in (Stilgoe et al., 2013a). E-PEACE has received almost no negative attention from the media or ENGOs, in stark contrast to the experiences of other projects such as SPICE and LOHAFEX which identified as geoengineering from the outset (Shackley, 2013). The experiment raises important questions over how governance could or should cover atmospheric research that has implications for SRM after it has taken place.

E-PEACE, led by PI Lynn Russell, and funded by the National Science Foundation and the Office of Naval Research, sought to test three hypotheses, the third of which being that it is important to have giant cloud condensation nuclei (CCN) to initiate drizzle in clouds (previously proposed by Bruce Albrecht) (Albrecht, 1989). This was tested by using artificial salt (similar to that used for snow machines) dispersed from an aeroplane directly into clouds in order to see whether conditions could be created which would initiate drizzle, as well as using smoke generators on board a ship. The results of the research found that there are many conditions in the real atmosphere that are not well represented in the models, suggesting that global models had been estimating incorrectly the effect of particles (Russell et al., 2013). The results also found that "smoke generators on board smaller ships could provide a net cooling effect, which could be used to offset some of the warming caused by ship CO_2 emissions" (ibid.); thereby effectively crossing over into SRM research.

While E-PEACE, like other atmospheric experiments, has had relevance for SRM (Keith et al., 2014), it complied with all necessary legal obligations asked of it, and was carried out off the coast of Monterey, with small and expected environmental impacts. The high level assessment process advocated and explored in the SPICE experiment was not conducted nor called for, as the project never set out to study SRM. Nevertheless, the results had relevance for geoengineering and this reinforces the importance of intentionality within SRM field work (Robock, 2012; Stilgoe et al., 2013a, b).

E-PEACE has drawn some criticism in the media, with questions of whether it signifies "geoengineering research by another name" (Black, 2012). Some have gone as far as to argue that E-PEACE indicates "unregulated (SRM) outdoor experimentation has already begun" (Robock, 2012). Despite these criticisms, E-PEACE has been subject to significantly less negative attention than research that self-identifies as geoengineering,[3] with most criticism hailing from scientific circles rather than the wider public.

Research in cloud-aerosol interaction and other atmospheric experimentation, such as weather modification, is normal practice in many different parts of the world and not considered as controversial as SRM. The challenge for governance is to ensure that SRM research is not able to sidestep the obligations which many are calling for by simply classifying itself in another discipline. To say that this was E-PEACE's intention would be unfair to the researchers who were pursuing a legitimate scientific study, but their results are a warning sign for future governance highlighting the relative ease with which researchers could perhaps avoid SRM assessment processes. The struggle for governance is how to cover research which could transition into SRM (intentionally or not) at a later stage.

SPICE

The Stratospheric Particle Injection for Climate Engineering (SPICE) field experiment was part of a larger research project to examine the feasibility and likely effects of SRM deployment,[4] focusing on lab-based evaluation of candidate particles, climate modelling, and potential delivery systems. The project had a small outdoor fieldwork component, with a proposed field experiment to test the feasibility of a balloon deployment system scheduled to take place in Norfolk in 2012. A scaled-down version of a proposed balloon deployment method would be examined, spraying a relatively small amount of water from a 1km pipe (Pidgeon et al., 2013). While there was no serious environmental risk or any climatic impacts identified with the experiment, the team was keen to engage in wider social concerns, employing social scientists to conduct stakeholder engagement and exploring different methods of assessment (Corner et al., 2012). It was this engagement, amongst other issues over intellectual property, which ultimately contributed to the field experiment being cancelled when the researchers were left feeling uncomfortable with what their experiment could signify (Watson, 2014). SPICE demonstrates the challenges for governance in dictating how scientists should scope their assessment process and how to define what is considered acceptable outdoor SRM research.

Seeking to engage with the social concerns raised by many (Corner and Pidgeon, 2010), the SPICE team, with the assistance of scientific and NGO representatives, established a stage-gate assessment process (Corner et al., 2012). It was decided the experiment should involve both public and media engagement, and be "forward looking in scope" as well as tackling the environmental and legal aspects (Watson, 2014). As the experiment was not planning on conducting any actual geoengineering, the environmental and legal aspects were relatively simple processes in comparison to ensuring the project was "forward looking" (ibid.). Matt Watson, the project PI, hoped the stage-gate process would "stimulate debate"; to achieve this they sought to ensure their research was transparent, acknowledging there were areas they were uncomfortable with (ibid.).

The completed assessment conducted by the SPICE team resulted in an approximately 100-page document, half of which focused on environmental safety and the scientific technicalities (ibid.). The rest was described by Watson as more prospective, addressing issues such as: how people could use the technology, concepts, and risk of technological lock-in, where the field could be in 20 years, and how to engage with the media and stakeholders (ibid.). As part of this process, the project team completed an environmental impact assessment (EIA), or as Watson remarked, "an exercise in why it wasn't necessary to carry out an EIA" (ibid.), alluding to the negligible environmental impacts predicted. Stakeholder engagement was conducted both locally and nationally, with results indicating that "almost all participants were willing to entertain the notion that the test-bed as an engineering test – a research opportunity – should be pursued" (Parkhill and Pidgeon, 2011). The team also held a press briefing, publicly launching the proposed experiment at the national science festival in Bradford.

While the public engagement resulted in comparatively indifferent opinions on the experiment, the feedback from ENGOs was less than favourable. In retrospect, Watson has argued that this aspect of the engagement could have been better handled (Watson, 2014). The ETC Group in particular was a vehement campaigner against the field test. Their "Hands Off Mother Earth" campaign fought hard for a cancellation of the project characterising it as the "Trojan Hose" – a distraction from important mitigation policies and a proponent of SRM deployment (ETC Group, 2011). A number of environmental groups joined their campaign against the project, arguing that it was likely to contravene a decision by the UN Convention on Biological Diversity (CBD) to permit only "small-scale" field trials. Penning a letter to the UK government arguing this point, the ETC garnered the co-signing of 50 other ENGOs (ibid.). Despite the experiment predicting the smallest environmental impact of the three case studies, the negative campaigning witnessed by SPICE vastly overshadows that experienced by the other two.

The project was cancelled in May 2012 due largely to issues of governance and ownership, after encountering issues over intellectual property rights. Watson claims the ENGOs depicted the cancellation as SPICE having given into their pressure, arguing that this was an unfair representation, as they had always planned for the stage-gate process to be a learning experience: "What is really frustrating is that these differing representations undermine the fact that we spent a lot of time agonising over these issues and wanting to ensure the experiment was done right" (Watson, 2014). SPICE had sought to carry out an experiment that was reflexive and forward looking, engaged with difficult social and ethical questions. The results of this process left the team feeling "uncomfortable" with what the experiment could signify, and this, at least in part, led to the cancellation (ibid.).

While SPICE was transparent and represented no environmental risk it was still subject to substantial criticism, scepticism, and outright opposition. Unlike the first two case studies, which for various reasons did not seek wide-scale stakeholder engagement, SPICE was faced with the greatest level of opposition because of direct involvement in it. The cancellation of the field trial by the SPICE team has been considered by some as an example of "responsible self-governance in the absence of governmental oversight" (Long et al., 2012). As not all previous examples of outdoor research have been interested in taking on the same issues, it is for those who devise future governance frameworks to decide what the suitable parameters for any SRM assessment process may be.

Conclusion

These case studies demonstrate that researchers interested in exploring the myriad of social issues identified with SRM fieldwork have struggled to find an appropriate assessment process, while those who are less interested have carried out their research under little scrutiny. The E-PEACE experiment highlights the difficult challenge for SRM governance, no matter what form it adopts, in ensuring scientists (whether purposely or not) do not avoid wider concerns by how they define their experiments. In a similar vein, the solar radiation experiment witnessed in Russia highlights another issue not easily resolved: how can research taking place in countries that pay little heed to international regulation be governed? These first two case studies highlight the ways in which scientists can currently avoid engaging with the concerns addressed by others. The SPICE experiment raises a different question: how should researchers interested in pursuing responsible and transparent fieldwork scope their assessment process? SPICE sought to take on this question and was left feeling that the experiment should not go ahead. With a range of potential future SRM experiments proposed (Keith et al., 2014), there is a need for future governance frameworks to meet these challenges.

First published online as an opinion article in 2015.

Notes

1 See Izrael's biography at: http://heartland.org/yuri-izrael
2 For example, see the Climate Geoengineering Governance project at http://geoengineering-governance-research.org/
3 For examples, see section on the SPICE Project (later in this study) and LOHAFEX, an ocean iron fertilisation experiment (see Schiermeier, 2009).
4 See webpage of the SPICE project at: http://www.spice.ac.uk/about-us/aims-and-background/

16

VILLAGE SCIENCE MEETS GLOBAL DISCOURSE

The Haida Salmon Restoration Corporation's ocean iron fertilisation experiment

Holly Jean Buck

Haida Gwaii is a temperate rainforest archipelago in the North Pacific, six hours by ferry from Prince Rupert, British Columbia. It is home to about 5,000 people. Half are citizens of the Haida Nation, which is made up of two communities with their own governing councils. Skidegate band sits in the south, and 100 kilometres north is the band of Old Massett. Quiet and remote, Haida Gwaii made occasional news over the past three decades due to its activism for indigenous care of forest resources. Yet, on October 15, 2012, Haida Gwaii became the centre of a different kind of media maelstrom when British newspaper *The Guardian* released the story, "World's biggest geoengineering experiment 'violates' UN rules".

In the summer of 2012, the Haida Salmon Restoration Corporation (HSRC), a private partnership ocean stewardship company, released 120 tons of iron sulphate and iron oxide into an ocean eddy centred 400 kilometres west of Haida Gwaii and monitored the resultant plankton bloom with a fleet of high-tech ocean gliders and drifters. Their research question: "Does adding a trace amount of iron to an HNLC[1] ocean eddy located in a known salmon migration route cause phytoplankton to grow, and if so, what are the resulting environmental benefits or costs?".[2] Initially, the project was linked with a self-styled eco-entrepreneur, Russ George,[3] who had a history of contentious carbon credit start-ups.

HSRC's project has been variously pointed to as a demonstration of why geoengineering needs governance protocols, a justification for a ban on geoengineering research, and, less frequently, a tale about a community on the front lines of ecological change trying to proactively restore ocean ecosystems with meagre resources. The purpose of this chapter is not to judge the legality or science of the project, but to explain how and why the story developed and unfolded as it did, and examine what can be learned. The three themes this study will explore are the tension between citizen/village-scale science and institutional science, the media response to the event, and the slippery definition of geoengineering.

Salmon, carbon sequestration, and media: an accounting of events

To begin, we will look at the motivations and genesis of the project, which was designed to respond to the problem of salmon decline and to possibly become funded by carbon credits. Overall salmon decline has scientists puzzled, and salmon runs fluctuate wildly (Parsons and Whitney, 2012). The Haida of Old Massett have run a salmon hatchery on the Yakoun river for 40 years and know that salmon are spawning and leaving Haida Gwaii – they just aren't coming back in corresponding numbers. This led the HSRC to wonder: what if the problem is at sea?

Global phytoplankton biomass has been declining over the past century at a rate of about 1 per cent of the global median per year, according to a 2010 paper in *Nature* which sparked intense debate (Boyce et al., 2010). Winds deliver the Pacific micronutrients from east Asia, but HSRC's thinking is that land use change and climate change can mean less dust on the wind, implying that the bottom of the food chain is undernourished. Some scientists have hypothesised that iron-rich ash scattered by a 2008 Aleutian volcano eruption factored into the massive 2010 salmon run (Parsons and Whitney, 2012), though definitively proving correlations between ash, plankton, and salmon is difficult at best. But this idea about insufficient dust and insufficient plankton is significant in that it implies that HSRC's project mimics a natural process and that the natural process has been disrupted by humans already, perhaps making remediation a human responsibility.

Part of the contention around the project was that HSRC was hoping to sell carbon credits to make it economically sustainable. Right now, no formal carbon market is administering, selling, or validating credits from ocean fertilisation (Buesseler et al., 2008). Yet whether carbon markets could, or should, play this role is central in ocean fertilisation debates. Phytoplankton is responsible for half of the organic matter production on earth (Boyce et al., 2010), and the equivalent of about a quarter of annual anthropogenic CO_2 emissions is absorbed and stored by the ocean (Rau, 2014). If a plankton bloom can sequester some excess carbon, feed some salmon, and be funded by the carbon market, that seems like a triple win.

To understand how HSRC could consider carbon credits as a funding source despite the many uncertainties associated with ocean fertilisation (see Cao and Caldeira, 2010; Smetacek, 2012; Secretariat of the CBD, 2012), consider that Old Massett has long been interested in restoration and payment for *forest* ecosystem services. Like many environmentalists and green entrepreneurs, Old Massett economic development officer John Disney (2013) also believes that "humans must put a value on the environment to ensure its survival", and that this value must become a fundamental component of the way our economy works. Payment for *ocean* ecosystem services also could make the project self-sustaining – good science requires long-term monitoring. However, HSRC was able to take the salmon project forward with village funds: Old Massett voted to spend $2.5 million for salmon restoration, with about 200 of 700 citizens participating and the majority in favour (McKnight, 2013).

Press: the world is "alerted"

How, then, did this project come to be defined as geoengineering? While press coverage portrayed HSRC's project as geoengineering, HSRC did not consider it as such.

"We always thought of this as a village project whereby benefits, environmental, financial and social, would accrue to the village and citizens of Haida Gwaii", says director and operations officer Jason McNamee. As a village project, it had to obtain buy-in from the village. Critics claim that villagers were "duped" or didn't understand the project. In any case, the research was not a secret: many locals were appraised, Environment Canada staff knew about the idea since 2011 (McKnight, 2013), and the HSRC published project information on its website. John Disney (2013) explains the project's development: "In the seven years of preparation for the project with all the legislative chores, the financial planning, the computer modelling, the legal investigations, the endless discussions with the team and with my council, I never once heard the term 'geoengineering'. Nor did I ever hear terms such as 'controlling' the climate or 'managing' the salmon stocks".[4]

Perhaps the project became geoengineering on October 15, 2012, when *The Guardian* presented it as such. In October 2012, the ETC Group – a technology watchdog group with nine members and a million-dollar budget – "contacted international press outlets to alert them" about the project (ETC, 2012). Criticism erupted. The "revelation" of the HSRC project occurred during the UN's Convention on Biological Diversity in Hyderabad, India (CBD COP 11), in which the ETC Group was pushing for an enforceable test ban on geoengineering rather than the current non-binding moratorium.[5] Political analyst Josh Horton commented that the ETC Group and sympathetic reporters "have orchestrated a mini-scandal timed to coincide with deliberations" (Horton, 2012a). This would have been a familiar strategy, as ETC had cast another ocean fertilisation experiment, LOHAFEX, as a violation of the moratorium in 2009 (Strong et al., 2009; Horton, 2012b).

Meanwhile, Haida Gwaii residents were coping first with the international press, and then an offshore earthquake. ETC claimed, "As Haida villagers headed for the hills amid tsunami warnings, they were still experiencing the aftershocks of the media storm of the previous fortnight" (ETC, 2012). Controversy arose within Haida Gwaii as well: the project was Old Massett's, but the negative press did not differentiate between the two Haida bands, placing new strains on the Skidegate/Old Massett relationship. The project was also not in conversation with the existing collaborative multi-year marine use planning efforts led by the Council of the Haida Nation (Jones et al., 2010), and the latter issued a press release stating its non-involvement (CHN, 2012). At least one resident claimed that the research was about selling carbon credits and not about salmon (Lavoie, 2012), and others called for an apology.

In the international sphere, some geoengineering researchers slammed the project – David Keith called it "hype masquerading as science" (Hume, 2012) – and parties to the London Convention/London Protocol[6] released a statement of condemnation in early November 2012. Joe Spears, legal counsel for the salmon corporation and village of Old Massett, called the condemnation "a clash between big science and big [non-government organisations], and village science and indigenous peoples" (Lavoie, 2012).[7] A legal case ensued between the HSRC and Environment Canada. Meanwhile, in autumn 2013, pink salmon made a strong return to BC waters (Hume, 2013),[8] an event that HSRC plans to analyse further as it continues its work. Perhaps the project will cease to be seen as "geoengineering" in the future, as HSRC's data continues to be shared. In this case, "geoengineering" is what happened when actors collided.

Ocean fertilisation and geoengineering: further exploration

The complex context of the Haida salmon restoration project suggests that we might need to think a little differently about climate engineering governance. In this analysis section, we will raise three questions.

First, who has the right or responsibility to act when faced with ecological decline? Disney comments, "It seemed appropriate for Old Massett to take the first steps to reclaim their stewardship role by working in an area that, before contact, would have been their responsibility". Colonialism is part of this context: the Haida population decreased after contact in the 1800s by 90 to 97 per cent (Martineau, 1999). Haida Gwaii's abundant resources have historically been exploited for some external interest's gains. Economic hardship is also present: Old Massett's high unemployment (~70 per cent) is mitigated somewhat by residents' abilities to fish and gather food, which makes functioning ecosystems and access to local foods like salmon even more important. In this context, payment for ecosystem services takes on a different rationale than the get-rich-quick exuberance of financialised carbon; it seems like a responsible way to generate income. Another consideration in terms of responsibility to act is that Old Massett never signed a treaty ceding rights and title to their traditional territory, and could be considered the legitimate authority in its traditional lands. HSRC applied for and was granted three research permits from Old Massett's governing body, the Old Massett Village Council. While indigenous peoples have certain rights and responsibilities regarding their environment, Whyte argues that geoengineering governance models are generally silent about indigenous peoples or conceive of community members "only as citizens of nation states or as groups that have special rights under the constitution of the nation state that dominates them". He recommends that early SRM research governance models "articulate Indigenous peoples as sovereign peoples in relation to NGOs, private companies, scientific advisory committees, supranational organizations, as well as federal or state agencies of nation states" (Whyte, 2012). While specific to solar geoengineering, the advice is certainly applicable in this case.

Second, who decides what is legitimate science? Does the legitimacy come from the actors, the experimental design, or the funders? The case brings up questions of how established institutions keep control of "science" in the 21st century, where both information and equipment are readily available to the non-institutionalised. Perhaps new institutions, like the Haida Ocean Center of Excellence imagined by McNamee and Old Massett, where people could study environmental change in the North Pacific with open-source software and equipment, can be a forum for engagement. This topic is quite relevant for geoengineering governance, which often assumes an agreement between parties who are all already in the room, acting within professional scientist norms (and managing liability issues).

Finally, how do we disentangle "geoengineering governance" from environmental governance – or environmental care – more broadly? Disney's (2013) broader vision includes illustrating how "a group of determined, smart and diverse-minded people can set up totally sustainable systems to satisfy their own energy, food, transportation, health, education, spiritual and cultural needs"; on-island wind power, local food systems, community exercise and redefining economics are all part of his vision. It is not possible to separate out "geoengineering" activities from these socio-ecological concerns; nor is it possible to cleave it from natural resource use and access, which are at the heart of this project.[9]

In conclusion, this case has pointed to the mounting set of problems with the umbrella term "geoengineering". As a linking term, "geoengineering" served to connect the salmon restoration project not just with solar radiation management, but with imaginaries of global control, fossil fuel industry corruption, conservative think tanks, and a whole web of signifiers that are unconnected with this specific project save the semantic link. In this case, it was useful for activists to link the project to solar radiation management and other contentious strategies. Yet it is absurd to link these techniques – with their varying scales, mechanisms, and motivations – and at the same time keep them separate from "usual" planetary-scale modifications, such as run-off from industrial agriculture or deep-sea trawling. The umbrella term is useful in that it invites comparison of different possible approaches to address climate change. Still, the evolution of the umbrella term "geoengineering" into something more coherent and analytically stable is probably due.

First published online as an opinion article in 2014.

Notes

1 High nutrient/low chlorophyll.
2 Written interview, September 2013.
3 Old Massett had a ten-year working relationship with Russ George from tree planting work. With previous ocean fertilisation start-up experience, he was a scientist and advisor on the HSRC project until they parted ways in May 2013.

4 On geoengineering, Disney comments that he rarely hears of any geoengineering scheme that would "fix the mounting list of problems emanating from our industrialization of the planet. Most of them scare me".

5 A 2010 CBD moratorium invites parties to consider a ban on geoengineering activities beyond small-scale research until there is regulation.

6 A 1972 / 1996 protocol to regulate marine pollution and dumping of waste at sea.

7 In March 2013, Environment Canada raided HSRCs offices and took data and materials; a lawsuit about this is pending.

8 McNamee also states that while the HSRC team can't conclusively express a causal link with the surprising returns, they "hope that through DNA and stable isotope chemistry analyses we will be able to better define the relationship" (interview).

9 Galaz (2012), observing that geoengineering governance has thus far exclusively emphasised creating international level mechanisms, recommends approaches that integrate Earth stewardship and geoengineering.

17

CAPTURING THE IMAGINATION

Prospects for direct air capture as a climate measure

Duncan McLaren

To avoid more than 2°C of warming by this century's end, atmospheric carbon concentrations may well have to be limited to around 350 parts per million (ppm). This will be no mean feat – current carbon concentrations already stand at over 390 ppm. Closing the gap between 350 ppm and a plausible mid-range scenario in which concentrations rise to 500 ppm would require withdrawals of carbon dioxide (CO_2) from the atmosphere (or "negative emissions") of around 24 gigatonnes of CO_2 annually for 50 years (McLaren, 2012a).

Large scale negative emissions arguably rely on the cost-effective development of direct air capture (DAC) technologies which could draw down CO_2 using chemicals that bind with it. The basic technology of DAC has been used for decades in submarines and, more recently, spacecraft. However, the viability of such systems changes when scaled up to the infrastructural levels necessary for large-scale negative emissions. Powering DAC could generate more carbon than it captures and will require effective linkages with the development of carbon storage. These issues are largely irrelevant for smaller and more contained activities underwater or in space but are critical when applied to an intended climate measure.

The first proposals for climate DAC were published in 1999 (Lackner et al., 1999); subsequently, it has come to be seriously considered as a supplement to climate mitigation. Interest has grown as the difficulties of accelerating carbon uptake by biological and oceanic capture options have become clearer (McLaren, 2012a; Wiltshire and Davies-Barnard, 2015), highlighted by the counterproductive effects of some biofuels policies (see for example Fargione et al., 2008). This case study examines the development of DAC as a climate response and the obstacles remaining to its deployment at scale.

How would DAC work?

At its simplest, DAC works by absorbing CO_2 from the air onto a chemical receptor (or "sorbent"), which has a strong attraction for the CO_2, in a similar way as a sponge absorbs water. The most commonly considered sorbents are alkaline compounds based on calcium or sodium which form carbonates when exposed to CO_2 and amine solutions (as used in many carbon capture and storage (CCS) demonstrations), although various other options are being explored (Jones, 2011). Alkaline sorbents typically require more energy than amines, but the technologies involve lower capital costs.

Even with an effective chemical sorbent, this leaves a number of challenges. First, to capture CO_2, air must somehow be moved over the sorbent. When cleaning with a sponge, you take the sponge to the spill, but in DAC, the CO_2 in the air is typically brought to the collector. Many current proposals seek to use the natural power of the wind to do this, but some would rely on powered fans or exploit the air currents created by cooling towers. Second, once the sorbent is saturated it stops catching CO_2 and must be "regenerated" in some way. You might wring out your sponge, set it somewhere warm to dry out or rinse out the dirt with running water. All three methods are being tried to regenerate DAC sorbents: "pressure swing" (squeezing out the CO_2), "temperature swing" (warming up the sorbent to release the CO_2), and "humidity swing" (washing out the CO_2). In all cases, the CO_2 must be released from the sorbent in a controlled environment, so it can be compressed and stored away. Third, either at the capture stage or the regeneration stage, the CO_2 can get mixed with other chemicals or gases and may require further purification as well as compression before it can be used or stored. Finally, all three steps require energy, which is the underlying limitation of large-scale DAC.

DAC as a climate response

DAC is not expected to be an alternative to conventional mitigation nor to point-source abatement through CCS. Rather, it is intended as a supplement to such responses (Jones, 2011), providing a means to reduce ambient atmospheric concentrations of CO_2 or to offset dispersed emissions that are difficult to mitigate directly, such as those from air travel. The big challenge is that CO_2 is very dilute in the air, which makes its capture energetically and financially difficult. But a growing body of research suggests a range of technical options and business or policy models that might make DAC viable (Keith, 2009b) (Table 17.1).

Given the slow progress achieved in the deployment of CCS as a climate response, it is important to understand why we might also seek to capture CO_2 from the air, rather than from the CO_2-rich flue gases emitted from power stations. Developers of DAC seek to optimise the efficiency of their processes at much lower capture rates than the 90 per cent typically expected of CCS on a power station. This cuts energy use *per unit* of CO_2 captured, but increases capital costs, as a larger facility would be required to capture the same *total* CO_2. Thus, there is good reason to develop the two approaches in parallel.

TABLE 17.1 Some leading proposals for direct air capture

Lead scientist	David Keith	Klaus Lackner*	Peter Eisenberger
Company	Carbon Engineering	Infinitree, formerly GRT & Kilimanjaro Energy	Global Thermostat
Funding	Private investors including Bill Gates	Venture capital	Private investors
Technology characteristics	Alkaline sorbent scrubbing; temperature swing regeneration; relatively low capital cost components but energy intensive – relying on stranded energy and waste heat.	Humidity swing amine sorbent; high capital, low energy cost. Relies on passive flow of air for capture and drying, but therefore consumes significant amounts of water.	Temperature swing amine; sorbent releases its CO_2 at a lower temperature than CCS amines; aim to use waste heat at ca. 90°C, from refineries, cement plants or smelters (Eisenberger et al., 2009); capital accounts for at least 40% of cost estimates.
Outputs	High purity CO_2	CO_2-enriched air	High purity CO_2
Development stage	Prototype demonstrated in 2012; small scale pilot plant (500t-CO_2 pa) under construction in 2015.	Lab-scale demonstration.	Small scale module (ca. 2t-CO_2/day) operating since 2013. Larger-scale demonstration plant planned for 2016.
Possible or published strategies for commercial development	Currently targeting direct production of synthetic fuel. Have considered integrating air capture into low-carbon liquid fuels markets (via EOR†); and fertilising algae for biofuel production.	Algal or greenhouse fertilisation; or with addition of purification and compression, chemical or food and drink uses of CO_2 might be initial customers.	Sale of CO_2 for EOR, algal fertilisation, or incorporation into cement or plastic; or for carbon tax credits.
Estimated costs	Long-term below $100/tonne-$CO_2$, with estimates of the pilot plant costs understood to be around $135/tonne.	Theoretical estimates of $25 – $40/tonne (Lackner, 2009; Lackner et al, 2012). But business plans imply $200 – $300 per tonne for compressed purified CO_2.	Latest claims approximately $50/tonne‡ (assuming use of gas for heating, but before compression and transport).

* As of 2015, Lackner is no longer personally involved in this attempted commercial deployment.

† Pumping captured CO_2 into oilfields to enhance oil recovery, and formally 'offsetting' it against the CO_2 produced when the oil is burnt, so as to market the resulting fuel as 'low-carbon'.

‡ Presented at Oxford Martin School, Greenhouse Gas Removal conference, October 2015.

It has also been suggested that rather than being tied to places where large amounts of CO_2 are emitted – as is the case with CCS – DAC could be cost-effectively co-located with CO_2 storage facilities or with low carbon energy resources which are "stranded" far from energy markets (and thus commercially unexploited), or developed on more remote, cheaper, less controversial sites (Keith, 2009b; Lackner et al., 2012). However, these apparent advantages are largely illusory. Specific site characteristics (such as low humidity) are likely more significant for DAC than co-location. And, in carbon terms, there is almost certainly more benefit in finding ways to directly replace fossil fuel use with heat that is currently wasted (in cooling towers for example) or solar power from remote deserts, rather than using such sources to power DAC. Moreover, remote locations can add to costs of construction and operation, while public concerns about CO_2 capture, transport, and storage are not limited to issues of direct exposure (McLaren, 2012b).

DAC options and obstacles to development

The credibility of the sector received a major boost in 2007 with the launch of the Virgin Earth Challenge (VEC)[1] which offers a $25 million prize for a scalable, sustainable, and commercially viable form of atmospheric CO_2 removal.[2] Among the front-runners for the VEC are companies based on the work of several of the technical leaders in DAC development: David Keith, Klaus Lackner, and Peter Eisenberger (see Table 17.1).[3]

While these and other VEC entrants are racing to pilot-scale demonstration, research into other methods and materials for carbon capture continues apace. Some promising possibilities such as metal oxide frameworks (MOFs) may prove suitable for air capture as well as flue-gas capture (D'Alessandro et al., 2010; Jones, 2011). Some MOFs appear to offer better absorbency without an equivalent increase in the energy costs of subsequent release. However, the behaviour of MOFs in real ambient conditions is as yet unknown.

Moreover, major obstacles remain in scaling up. To deliver significant negative emissions would require an industry as large in physical scale as the current oil and gas sector (McGlashan et al., 2012). The obstacles to such large-scale commercialisation include energy requirements and storage availability as well as social acceptability, cost, and funding mechanisms.

At scale, even relatively efficient forms of DAC would require very substantial amounts of energy. McLaren (2012a) compares the estimated energy requirements for capturing 24 gigatonnes of CO_2 a year with current world global energy consumption. Alkaline scrubbing methods would require the equivalent of an extra 60 per cent of total energy use, and even amine-based methods around 10 per cent – in addition to the massive efforts already foreseen to transform and extend existing energy systems. A breakthrough in nuclear fusion might make this plausible, but otherwise, deployment of DAC will likely compete for scarcer and more expensive energy.

DAC developers often assume that carbon storage will be developed for CCS and can be treated as a simple "bolt-on" to DAC. However, basic questions about storage location and availability, and regarding purification and compression before transmission and storage, have not been fully considered. While official upper estimates of global carbon storage (IPCC, 2005) would be more than adequate for DAC to achieve safe atmospheric levels of CO_2, at the more conservative end of the range, estimates of affordable storage capacity are below the likely aggregate requirements for negative emissions. Regional scarcity might also be significant. In some regions, concerns about safe and publicly acceptable storage availability have already begun to influence CCS policy (McLaren, 2012a).

The social acceptability of DAC is far from assured. Public opposition to carbon storage has hampered CCS development in several countries (McLaren, 2012b), while place-based concerns about energy infrastructure (such as wind farms) could be replicated in the context of proposals for large-scale deployment of air capture devices (see, for example, IPSOS Mori, 2010).

The cost estimates of promoters appear optimistic and cannot be verified without much greater transparency. Keith's cost estimates contrast remarkably well with the figure of $600/tonne or more suggested by the American Physical Society for a similar, but generic, system, in one of the few independent assessments of the likely costs of air capture (APS, 2011). Keith has suggested ways in which the APS could have overestimated (Carbon Engineering, 2011) but these seem unlikely to account for the whole discrepancy. Lackner's and Eisenberger's published cost estimates are also lower than figures estimated in recent theoretical assessments at $240/tonne (McGlashan et al., 2010) and $1,000/tonne (House et al., 2011). In practice it seems costs are likely to lie somewhere between the theoretical estimates and those of the promoters.

Thus, reaching commercial viability may require strategies that undermine the effectiveness of DAC as a climate measure. These might include algal fuels and enhanced oil recovery (EOR) through CO_2 injection into oilfields. The figure of $50/tonne is often cited as a level at which sale of captured CO_2 for EOR could be commercially viable. But the combustion of the additional oil produced through EOR releases more carbon dioxide than is captured in the EOR process. Overall system emissions are only reduced if less oil is produced elsewhere as a result of EOR or algal biofuel production, but this is at best uncertain in a market that is currently supply constrained. Marketing DAC-EOR oil as "lower-carbon oil" under national or state fuels standards systems (Carbon Engineering, 2013) might facilitate partial abatement, but at the risk of increasing the lock-in of existing fossil vehicle technologies. Moreover, the business models under development suggest that even the $50 level may remain elusive, as they often rely on potentially higher value but niche carbon utilisation markets (e.g. for chemical or food grade CO_2), or on regulated or voluntary carbon offset markets alongside EOR.

For most DAC proponents a functioning carbon market with high carbon prices is seen as ideal. Yet financing DAC through carbon markets – whether official or voluntary – is problematic for climate policy. DAC is at a stage at which it is not competitive given existing carbon prices. To incentivise development will likely require direct support or specific regulatory drivers. Yet if it becomes competitive, unless the overall market caps are reduced, DAC would then drive out, rather than supplement, other mitigation approaches – acting purely as an offset. This is particularly problematic if storage availability proves to be limited (McLaren, 2012a).

Prognosis and conclusions

Large scale DAC remains a "technological imaginary". It is no silver bullet for the climate problem, despite being almost certainly essential to long-term reductions of atmospheric CO_2. Even with further technological and process breakthroughs, it is likely to be expensive and slow to roll out. From an optimistic perspective, technical developments in DAC might feed back into more efficient CCS processes, unblocking wider deployment of that technology. On the other hand, this could be just another example – alongside business models predicated on EOR – of how DAC might further lock-in societal dependence on fossil fuels.

Given such risks, the case for DAC research being guided by the principles of responsible innovation (Stilgoe et al., 2013a) is strong. However, most research is being conducted in the USA, funded primarily by private venture capital. This makes for high levels of commercial confidentiality and tight control over intellectual property while risking promoter bias. Obtaining investment demands financial projections in which profitability appears credible, encouraging promoters to use optimistic technical assumptions. It also requires positive publicity and confidence in the prospects of the technology and its possible market, so a bubble of inflated expectations can easily result (Hansson, 2012; McLaren, 2012a).

Yet belief in the future availability of relatively cheap DAC could fuel reluctance to take timely action on mitigation. Many modelling studies suggest the presence of negative emissions technologies leads to an allegedly "economically efficient" delay in mitigation (for example, see Keith et al., 2006). If DAC captures imaginations too well, it could increase the climate risks we face from possible tipping points in the system.

Policymakers must take care: with the right imagination and sound incentives DAC could accelerate mitigation and widen future options – done badly, policy for DAC could exacerbate lock-in and undermine progress.

First published online as an opinion article in 2014.

Notes

1 *Virgin Earth Challenge*. Accessed February 2014 at: http://www.virginearth.com/
2 Cynics might suggest that VEC's sponsor, Sir Richard Branson, is protecting his aviation interests, but the prize and involvement of high profile judges such as Al Gore and James Hansen has inevitably raised the profile of serious contenders.
3 The VEC and its candidates provide high-profile examples of the most commercially advanced technologies and options that are on the table; and concrete illustrations of funding sources, cost, technology, and claims made by developers.

PART IV

Existing institutions and emerging frameworks

Introduction

While the last section demonstrated that technological progress towards the development of deployable geoengineering technologies has been very limited, the potential for future manifestations of these technological ideas to substantively influence the Earth's climate remains significant. But as the concept of geoengineering gradually permeates the public and politic mainstream, it is not emerging into a vacuum. Numerous national and international institutions for managing diverse aspects of the environment have been built over the last half century, including ones specifically designed to tackle climate change. This section of the book explores how geoengineering is currently interacting with, and might or might not in the future fit within, these existing frameworks.

Chapter 18 launches this section with Arthur Petersen leveraging his personal experience with the Intergovernmental Panel on Climate Change (IPCC) – the international partnership between scientists and governments to produce a scientific review of climate change every handful of years – to examine how the IPCC has addressed geoengineering throughout its evolution to the Fifth Assessment Report (AR5). This chapter shows that what geoengineering approaches are included in AR5, as well as how their benefits and risks are communicated, were the result of interplay between climate negotiators at the UNFCCC and scientists in Working Groups I (climate physical science) and III (mitigation pathways) of the IPCC. The reader would also benefit from examinations of historic and current IPCC-UNFCCC interfaces that provide further context to Petersen's insider perspective (Hulme and Mahony, 2010; Guillemot, 2017), as well as on how SRM and CDR might in the future be situated within the legal and policy framework of the Paris Agreement. (Craik and Burns, 2016 make an educated attempt, though they stress that neither suite has been formally adopted under UNFCCC auspices.) Since the time of writing, SRM has yet to receive further attention within the context of the Paris Agreement. CDR, on the other hand, has been the subject of criticism for the heavy role played by bioenergy carbon capture and storage (BECCS, a form of CDR) in AR5 WGIII modelling scenarios that lead to low climate targets (2°C and 1.5°C). Critics argue that modellers have allowed an unproven and controversial carbon removal approach to underpin the viability of the Paris Agreement's targets, raising questions of whether IPCC assessments are being increasingly structured by – and called upon to legitimate – political objectives, and of the role of scientists in negotiating this tension (Geden and Beck, 2014; Anderson and Peters, 2016; Beck and Mahony, 2017).

As IPCC reports are designed to inform the international climate negotiations, this provides an important foundation for the discussion that Jesse Reynolds picks up across the next two chapters. In Chapter 19, Reynolds provides a synoptic review of how existing international legal frameworks might apply to different geoengineering technologies; and in Chapter 20 he extends his analysis to argue why existing international environmental institutions should not be attempting to regulate solar geoengineering technologies today. Rounding out this section is Chapter 21, wherein Joshua Horton and colleagues examine examples of international liability regimes as potential precedents for developing a similar regime for prospective SRM geoengineering technologies. Indeed, there is a rich context of actors, issues, and laws whose piecemeal applicability to the debate can become suddenly and influentially relevant, and latent linkages to wider politics may be the most important part of tomorrow's landscape. As Reynolds points out, at the international level, concerns over biodiversity impacts (Convention on Biological Diversity) and the marine environment (London Convention and Protocol) already do. There may be some lessons here: a wide range of institutional forums with mandates and self-sustaining agendas exist and are likely to be applied proactively or retroactively to geoengineering approaches.

Reynolds's and Horton et al.'s reviews should also be read alongside other analyses that have helped to fill in understandings of the "regulatory gap" for geoengineering governance at the international level (Armeni and Redgwell, 2015; Armeni, 2015a) and at the national level (Hester, 2011, 2013; Armeni, 2015b, 2015c). Many early reviews and assessments of governing institutions focused on the constellation of UN environmental regimes, which has since been argued to have infrastructures and mandates more suited to constraining deployment than facilitating the marshalling of funding and coordinating research at the upstream stages (Armeni, 2015a; a point made also by Reynolds, as well as Ghosh in Chapter 31 of this volume). Adapted mechanisms at the national level, as well as non-legal, researcher-driven governance frameworks (see Kruger, Chapter 30; and Foley et al., Chapter 34) have been highlighted as more applicable at early and more explorative stages. We explore the latter, which has seen much growth in the literature, in Part VI.

18

THE EMERGENCE OF THE GEOENGINEERING DEBATE WITHIN THE IPCC

Arthur Petersen

The Intergovernmental Panel on Climate Change (IPCC) has some agenda-setting power for global climate policy. This explains recent worries about the fact that the governments had decided in 2009 that the IPCC's Fifth Assessment Report (AR5) was to explicitly address geoengineering options, which could then possibly legitimate the serious consideration of such options in global climate policy negotiations. Such worries, however, neglect two factors. First, the IPCC has a long history of dealing with geoengineering and, second, the IPCC performs its assessments without endorsing any options and being based on what is available in the primary literature. Still, there is no way to deny that the way the IPCC summarises the science does have an influence on how a particular subject is subsequently discussed in policymaking. For that reason, it is already interesting to look back at the emergence of the geoengineering debate within the IPCC.

From my analysis of IPCC reports, a few trends become clear. Geoengineering—in all of its forms and using the term "geoengineering" – was part of all the first four rounds of IPCC reports since 1996, at the level of both individual chapters and summaries for policymakers (SPMs). Geoengineering has also never been endorsed by the IPCC. However, in some of the IPCC reports further study of geoengineering options has been promoted, and the latest IPCC report (AR5, 2014) made it clear that reaching a two-degree target would in many scenarios entail large-scale afforestation and/or production of bioenergy with carbon dioxide capture and storage (BECCS).

From the First Assessment Report (1990) to the Fourth Assessment Report (2007)

In the First Assessment Report (FAR) of 1990, the reference made to geoengineering was limited to the discussion of large-scale reforestation and afforestation, with the summary for policymakers (SPM) of the FAR WGIII report explicitly mentioning these as being part of scenarios that would keep CO_2 concentrations within certain bounds. No other options for either carbon dioxide removal (CDR) or solar radiation management (SRM) were mentioned anywhere in the FAR, and the term "geoengineering" was not yet used by the IPCC.

The Second Assessment Report (SAR) of 1995 was the first IPCC report that assessed "geoengineering" options, which in the SAR WGII SPM were considered "likely to be ineffective, expensive to sustain, and/or to have serious environmental and other effects that are in many cases poorly understood". In chapter 25 on mitigation (still part of WGII at that time), geoengineering (both CDR and SRM) was discussed in a section on "concepts for counterbalancing climate change". Still, only SRM examples were given in the SPM.

Five years later, the Third Assessment Report (TAR) of 2001 mentioned geoengineering in its WGIII (mitigation) SPM under "gaps in knowledge": it argued that "some basic inquiry in the area of geo-engineering" was warranted. Interestingly, in contrast with the SAR, only CDR examples were given in the SPM this time.

In the Fourth Assessment Report (AR4) of 2007, the two examples mentioned in the SPM (of WGIII) were ocean fertilisation (CDR) and stratospheric aerosols (SRM), and geoengineering options were assessed to "remain largely speculative and unproven, and with the risk of unknown side-effects". It was also noted that "[r]eliable cost estimates for these options have not been published".

The Fifth Assessment Report (2014): Working Group I

It must be admitted that the assessment of geoengineering options in the Fifth Assessment Report (AR5) of 2014 has been the most extensive of all IPCC reports, mainly because much literature has appeared in the eight years before AR5. Still, even though an IPCC expert meeting on geoengineering held in 2011[1] had received some attention, it came as a surprise to some that the WGI SPM (which was approved by governments on September 27, 2013) contained a final paragraph, which read as follows:

> Methods that aim to deliberately alter the climate system to counter climate change, termed geoengineering, have been proposed. Limited evidence precludes a comprehensive quantitative assessment of both Solar Radiation Management (SRM) and Carbon Dioxide Removal (CDR) and their impact on the climate system. CDR methods have biogeochemical and technological limitations to their potential on a global scale. There is insufficient knowledge to quantify how much CO_2 emissions could be partially offset by CDR on a century timescale. Modelling indicates that SRM methods, if realizable, have the potential to substantially offset a global temperature rise, but they would also modify the global water cycle, and would not reduce ocean acidification. If SRM were terminated for any

reason, there is *high confidence* that global surface temperatures would rise very rapidly to values consistent with the greenhouse gas forcing. CDR and SRM methods carry side effects and long-term consequences on a global scale.

However, comparable text had been part of the first draft of chapter texts, which was circulated to experts for their review in December 2011. While the first draft of the SPM (of October 2012) – oddly enough – did not contain any reference to geoengineering, the paragraph quoted above did appear – in very comparable form – in the final draft that was distributed to governments in June 2013. And when the paragraph first came up for discussion in the plenary approval session in Stockholm in September 2013, no country raised its flag. Apparently, every government could live with the text as initially proposed by the authors, which was slightly amended in response to government review comments. Thus, there really was no debate on geoengineering in the IPCC WGI plenary in Stockholm in September 2013. And I must say that the paragraph's wording was very carefully crafted indeed.

The Fifth Assessment Report (2014): Working Group III

Similarly to the FAR of 1990, the AR5 WGIII SPM of 2014 emphasised again that for strong mitigation scenarios, large-scale afforestation could be needed to remove carbon from the atmosphere. But the main IPCC message pertaining to geoengineering in 2014 became that in many of the mitigation scenarios assessed, the geoengineering option of bioenergy production with carbon dioxide capture and storage (BECCS) had been used. The authors of AR5 WGIII SPM, however, did not use the term "geoengineering", preferring to refer explicitly to only these two geoengineering options. This was because only BECCS and afforestation had featured in their assessment of mitigation scenarios, and they were afraid that "geoengineering" might carry a negative association.

But on the third day of the WGIII plenary, it became clear that one country could not agree to the proposed text and the way geoengineering was framed in the SPM. The first intergovernmental geoengineering debate within the IPCC was born, only to be resolved after four sessions of a contact group that extended over the last three days of the plenary. I will here recount some of the discussions on geoengineering that were held in the plenary and in the contact group, which I co-chaired together with a delegate from Brazil.

In the plenary, it was pointed out by one country that the geoengineering options assessed by the IPCC were at odds with the UN Framework Convention on Climate Change and amounted to another invasion of developing countries by the developed countries. Furthermore, there is significant uncertainty pertaining to the effectiveness and side effects of geoengineering options. Policymakers must receive balanced information about these kinds of technologies and their limitations. This is a moral issue: the IPCC carries a special responsibility to give the most comprehensive and clear portrayal of uncertainties, risk, and limitations of geoengineering methods and technologies. The country further added that the IPCC should develop an ethical protocol for its own assessments.

In the contact group, having spent most of the time discussing how to prevent too much focus from the IPCC on mitigation scenarios that would keep the 2°C target within sight (the IPCC could then be seen to propose that this target would have to be met), there was wide agreement among countries to request that the authors include the following part of the approved WGI text on geoengineering in a footnote:

> According to WGI, CDR methods have biogeochemical and technological limitations to their potential on the global scale. There is insufficient knowledge to quantify how much CO_2 emissions could be partially offset by CDR on a century timescale. CDR methods carry side-effects and long-term consequences on a global scale.

Furthermore, the following text was added to the bold text of the paragraph on reaching the 2°C target through "overshoot scenarios" that involve negative emissions: "The availability and scale of these [afforestation and BECCS, acp] and other Carbon Dioxide Removal (CDR) technologies and methods are uncertain and CDR technologies and methods are, to varying degrees, associated with challenges and risks".

Thus I conclude that while governments were satisfied with the way geoengineering options were assessed in the final draft of AR5 WGI, they wanted more emphasis on the uncertainties and risks of large-scale afforestation and BECCS than was contained in the final draft of AR5 WGIII (even though these uncertainties and risks were already contained in one location in the text). By making use of some of the already approved text from WGI, it was not difficult to accommodate this wish from governments. Still, many issues, such as those pertaining to the governance of geoengineering and geoengineering research, were left untouched by the IPCC summaries, and it should be expected that were geoengineering to feature in future IPCC reports (e.g., a Special Report on Geoengineering), such issues will likely receive more attention.

First published online as an opinion article in 2014.

Note

1 See: http://www.ipcc-wg3.de/meetings/expert-meetings-and-workshops/em-geoengineering

19

THE INTERNATIONAL LEGAL FRAMEWORK FOR CLIMATE ENGINEERING

Jesse L. Reynolds

Several of the key, recurring questions which loom over climate engineering concern how countries would interact when some of them undertake or approve actions that might impact other countries. May a state intentionally alter the climate? What would its obligations be before, during, and after doing so? What if a potentially affected country protests or claims that it had been harmed? What if the implementing country believed that its existence was at risk due to impending climate change? What about private actors attempting climate engineering, perhaps for profit? Is there an existing legal instrument under which field tests with potential trans-boundary impacts could be regulated? Are countries obligated to research or implement climate engineering in order to prevent dangerous climate change? May states claim credit for greenhouse gases (GHG) removed from the atmosphere via carbon dioxide removal (CDR)?

Countries prevent and resolve international disputes through a variety of mechanisms. One particularly important mechanism is international law. This chapter describes some international law that is applicable to climate engineering, with a focus on international environmental law. It closes with a brief synthesis and some recommendations for future developments. First, though, it introduces international law and suggests why climate engineering is such a challenge for international environmental law and its scholars.

International law

International law is a collection of authoritative rules governing countries' actions, especially those that may impact other countries. That is, sovereign states are its subjects. With limited exceptions, international law governs neither the actions of individual persons nor those of national governments that have only domestic impacts. These governments may be, however, obligated to require, regulate, or prevent certain actions by their citizens and residents, although the states are not necessarily responsible for the actions of their persons.

Scholars offer a wide range of explanation for how and why international law operates, and this often shapes their conclusions as to what it can accomplish (Dunoff and Pollack, 2012). Some assert that it is an outgrowth of the shared values and intersubjective understandings of those individuals who craft it and that it thus carries strong normative power. Others claim that national leaders develop and implement international law in response to the domestic constituencies who support them. Finally, a third group argues that states with differing levels of power and capabilities rationally use international law to coordinate, cooperate, and coerce because it furthers their diverse interests.

The most important characteristic of international law is that there is neither a central legislator nor central enforcement. This is unlike the national law with which we are most familiar, which is developed through legislative processes and enforced through the state's threat of force. In contrast, international law is a set of promises, customary behaviours, and principles among purportedly equally sovereign states. These rules are of varying explicitness, detail, and "firmness", in the sense of their rhetorical strength, associated expectations, and possible consequences of their violation. Although these consequences are sometimes explicit in a treaty, most often international law is enforced in three general, indirect ways (Guzman, 2008). A victim country might reciprocate with the same violation back at the violator. States may also retaliate in other, unrelated areas. Finally, the violator frequently suffers in its reputation, and states are consequently less likely to engage with it in ways that would have been beneficial. Notably, enforcing international law is often costly for the enforcers, compounding the challenge.

International law traditionally has three primary sources. Treaties are explicit agreements among states that choose to participate. Most treaties (or similar terms, such as agreements or conventions) are between two countries, although some have many participants, called parties. Customary international law is what all countries consistently do out of an apparent sense of legal obligation, and applies to all states who do not explicitly object. Finally, general principles are the guiding ideas upon which treaties and customs are based, but are not themselves binding on their own. The precise substance of customs and principles are not centrally codified and thus sometimes disputed. Beyond these, the rulings of international tribunals and intergovernmental organisations have become important secondary sources within the international legal system.

Regardless, trying to make a sharp distinction between binding and nonbinding international law is mostly unproductive, even though certain components of it are clearly intended to be one or the other. Instead, there is something of a gradient. Furthermore, countries, particularly the powerful ones, sometimes violate explicitly binding agreements with little consequence, especially if there is a widely shared sense that the action was justified. Likewise, other countries, particularly the weak ones, sometimes face sanction for actions that are not contrary to international law. Although this implies that politics trumps international law at the end of the day, the latter still has an impact by altering the incentives that states face. Indeed, countries generally abide by international law. This can be explained variously by its genuine effectiveness, its ambiguity, or its mere embodiment of what countries would have done in its absence.

The international law of the environment is relevant when a state's actions pose risks to the environment of other states or of areas beyond national jurisdiction and control, such as the high seas, Antarctica, or outer space. For the most part, international environmental law is anthropocentric, in that it protects the environment for people's health and for their natural resources (Birnie et al., 2009, p. 7). Notably, it is intertwined with efforts to overcome uneven economic development.[1] That is, all countries want their own environments to be clean, but there are divisions of international priorities: wealthy states generally emphasise global environmental protection, while the poorer ones wish to develop economically and are concerned that stringent international environmental law could interfere with this (Najam, 2005).

As a final note, international law as described arose when countries were considered the exclusive actors in the international arena. In recent decades, transnational non-state actors have become increasingly important – or at least recognised as such – both as sources and subjects of a more broadly defined system of international law (Biermann and Pattberg, 2008). Indeed, so-called "global governance" instruments and institutions that rely less on states than traditional law may be more effective in regulating transnational non-state actors.

The challenges of climate engineering

Climate engineering presents difficulties for international environmental law and its scholars. To some degree, this is due to its novelty: climate engineering proposals have been seriously discussed for only a decade or so. This situation is frequently seen with new technologies, as international law moves slowly by design. In these cases, scholars and practitioners are forced to interpret legal instruments that were developed for decidedly different purposes.

Climate engineering is especially challenging because it presents three novel dynamics for international environmental law. First, all climate engineering approaches – both CDR and solar climate engineering (solar radiation management, or SRM) – could both prevent and cause environmental harm. Removing GHGs or increasing albedo could lower climate change risks while simultaneously creating new risks. For example, SRM would unevenly compensate the temperature and precipitation anomalies of climate change, CDR methods may alter ecosystems, crowd out food production, and create new industries of massive scales (National Research Council 2015a; National Research Council., 2015b). In short, these are proposed interventions to protect humans and the environment that may also harm humans and the environment. Indeed, both climate change (or GHGs) and climate engineering often satisfy the definitions of "pollution", "damage", or "adverse effects" that environmental treaties try to prevent and reduce (CLRTAP, 1979; UNCLOS, 1982; Vienna Convention for the Protection of the Ozone Layer, 1985; Madrid Protocol, 1991; UNFCCC, 1992; CBD, 1992; OSPAR, 1992). It is often unclear how international environmental law should balance such tension.

Second, SRM climate engineering does not fit the mould of typical environmental problem structures, which can usually be described economically as negative externalities or collective action problems with the environment as their medium. In the former, some actors engage in activities that are beneficial for them but have negative environmental consequences for others that are not taken into consideration by the former. In the latter, all would benefit if each were to take some action, but they would individually benefit more by not doing so and "free riding" on others' efforts. In contrast, SRM presently appears to offer a *positive* externality through the reduction of climate risks whose value would not be fully captured by the implementing actor, coupled with some risks for others.[2] Further, instead of a need for collective action that brings a free rider problem, SRM calls for collective restraint that brings a "free driver" problem (Weitzman, 2015). Therefore, although SRM would to some extent instigate traditional environmental law mechanisms such as preventing, remedying, and possibly compensating for harm, it appears that its research and possible implementation would primarily generate challenges such as coordination, mutual restraint, and prevention of misuse (Bodansky, 2012).

Finally, I suggest that there is a cultural barrier as well. The contemporary awareness of environmental degradation that arose in the 1960s and later provided a cultural foundation for modern environmental law is, at its core, a realisation that we have not adequately accounted for the full environmental consequences of our actions. This usually includes a belief that certain large-scale, high technology endeavours attempted to intervene in nature in ways that were dangerous and insufficiently understood. Most of the environmental movement has responded by calling for greater humility, increased scepticism of our knowledge and technology, and placing the natural world more centrally in our decision-making processes and value systems. In this context, climate engineering "runs afoul of almost every major trend in contemporary environmentalism" (Michaelson, 1998, p. 81). To the extent that this is the narrative behind the rise of environmental law, a logical reaction has been to see climate engineering not as a potential means to reduce net climate risks but instead as the latest in a series of hubristic technological threats to a fragile global environment.

Applicable existing international law

Here, I briefly review the most relevant existing international law in the context of climate engineering. Unsurprisingly, most of this falls under the rubric of international environmental law, although other domains will be briefly touched upon at the section's end. Unless otherwise stated, these instruments and provisions apply to all climate engineering techniques that would pose trans-boundary risks. However, some proposed methods – especially some within SRM – are more likely to do so.

International environmental law

International environmental law is the logical starting point for considering how international law may be able to prevent and resolve disputes arising from climate engineering. International environmental law is not a distinct domain but instead merely the subset of international law that relates to how states may impact each other via the environment. Although what is and is not an environmental matter is unclear (e.g., is a liability for harm from space activities an environmental issue?), this need not be resolved here.

As a starting point, states' sovereignty means that they are free to govern their people and to manage their resources within their territory as they deem appropriate, provided that such actions do not harm other countries (Rio Declaration, 1992). Per customary law, if an activity poses a risk of significant trans-boundary harm – including a high chance of typical harm and a low chance of "disastrous" harm – then the country of origin is obligated to take appropriate measures to prevent or reduce the harm; to review and (if appropriate) to authorise risky activities; to assess potential environmental impacts; to notify, consult, and cooperate with those countries likely to be affected; to notify the likely affected public; to develop plans in case of an emergency; and to monitor an activity's ongoing effects (International Law Commission, 2001a). In other words, the source state is to act with due diligence. Importantly, this is not to be done solely to minimise trans-boundary environmental harm but instead to equitably balance states' interests (including the benefits, importance, and risks of the activity), those of available alternatives, and the costs of prevention. If an incident has caused or is likely to cause trans-boundary harm from a hazardous activity, the source state should notify, consult with, and cooperate with the likely affected countries in order to take appropriate response measures, while the likely affected countries are to take all feasible measures to mitigate the damage (International Law Commission, 2006). Afterwards, those states that have caused trans-boundary harm through an action that was contrary to international law must stop the activity; assure that it will not reoccur; make reparations for the harm through restitution, prompt and adequate compensation (possibly by strict liability on the operators of hazardous activities), and satisfaction such as an apology; and provide access to legal remedies for victims (ibid.; International Law Commission, 2001b).

Several environmental agreements would be relevant in the case of climate engineering. Only a handful of treaties and treaty systems are discussed here; others would be applicable only in limited geographical areas and/or with particular climate engineering methods (Bodle et al., 2014; Reynolds, 2014a). The most important is the UN Framework Convention on Climate Change which now includes essentially all countries as parties (UNFCCC, 1992). Its objective of stabilising GHGs at safe levels and its binding commitments clearly indicate that CDR lies within its purview. Among the commitments are two that call for the enhancement of sinks and reservoirs. The UNFCCC's Kyoto Protocol is more explicit, requiring its parties to research and promote "carbon dioxide sequestration technologies and ... advanced and innovative environmentally sound technologies" (Kyoto Protocol, 1997, art. 2.1(a)(iv)). The questions as to whether particular CDR methods could be included towards a country's accounting of its net GHG emissions and whether they could be eligible for credit under international emission trading systems are important yet remain unresolved. The debates concerning the effects of forests, agriculture, and land use on GHG concentrations have dragged on for decades due in part to the complexity and uncertainty of their net long-term impacts; CDR methods will likely face a similarly difficult path. More recently, the Paris Agreement more explicitly points towards CDR in its goal "to achieve a balance between anthropogenic emissions by sources and removals by sinks of greenhouse gases" (Paris Agreement, 2015, art. 4.1).

The relationship between the UNFCCC and SRM is less clear. On one hand, these methods would not contribute toward its objective of GHG stabilisation. On the other hand, there are several references among its principles, priorities, and commitments that imply at least the consideration of SRM, perhaps through research. For example, the document's aspirational language calls for the prevention of dangerous climate change in a rapid and inexpensive manner "so as to ensure global benefits at the lowest possible cost" (art. 3.3), for anthropocentric reasons, and in balance with objectives such as economic development and food production. SRM may allow this to be done. Several commitments are to undertake research and to develop and diffuse new technologies in order to reduce uncertainty, including that of "various response strategies" (arts. 4.1(g) and (h)). Nevertheless, the mandate for the UNFCCC is unclear with regard to SRM, and whether its institutions will address the matter is ultimately a political decision (Reynolds, 2018).

The Convention on Biological Diversity (CBD, 1992) may be the most important general environmental treaty due to its broad provisions to protect biodiversity – which is affected by many large-scale human activities – and to its near-universal participation.[3] Among other things, its parties commit to several procedural duties concerning activities that are likely to have "significant adverse impacts" on biodiversity, which some climate engineering methods would (art. 7(c)). Perhaps more importantly, its Conference of Parties has taken an interest, agreeing to three statements regarding climate engineering (CBD, 2008, 2010a, 2012, 2016). These are the only statements on climate engineering in general that originated in a near-universal international legal forum. The 2010 one is a non-binding statement of caution, asking the parties to refrain from climate engineering that may affect biodiversity until there is the scientific basis for such work and "appropriate consideration of the associated risks". This request is to continue "in the absence of science based, global, transparent and effective control and regulatory mechanisms". It makes an exception for small-scale scientific activities. The 2016 decision reaffirms previous ones while calling for more research to improve understanding of climate engineering's potential impacts on biodiversity.

The Environmental Modification Convention is a less well-known multilateral agreement that prohibits the military application of weather modification methods (ENMOD, 1976). Its definition of "environmental modification techniques" would include most proposed large-scale forms of climate engineering and its parties may not use these for any "military or any other hostile use" (arts. I.1, II). Notably, the agreement explicitly "shall not hinder the use of environmental modification techniques for peaceful purposes" (art. III.1) and encourages peaceful applications of environmental modification. Although ENMOD includes most industrialised countries among its parties, it has no supporting infrastructure and is essentially dormant.[4]

The comprehensive UN Convention on the Law of the Sea (UNCLOS, 1982), with near-universal participation, would govern climate engineering activities that take place at sea or that would affect the marine environment.[5] Under it, states' obligation to protect the marine environment is without qualification. As noted in the previous section, its definition of "pollution" that states are obligated to "prevent, reduce, and control" includes climate change, GHGs, and climate engineering activities that are likely to be harmful (art. 1.1(4)). UNCLOS strongly supports scientific research provided that, among other things, it does not interfere with other states' legitimate uses of the sea and it is consistent with protection of the marine environment. The seas are divided into three zones, in which the first 12 miles are the territorial waters of the coastal states, up to 200 miles are the quasi-territorial "exclusive economic zone", and beyond that are the high seas, without national jurisdiction. Ships are the responsibility of their flag state, whose national laws apply to their crews.

The CDR method of ocean fertilisation warrants particular attention. It is the exception to the general rule that CDR would present well-characterised, low environmental risks and can mostly be regulated by domestic law. It is also the only (thus far) potentially high-risk climate engineering method to be repeatedly tested in the open environment. These outdoor experiments were conducted by universities and other public research institutions during the 1990s and 2000s. However, in reaction to private actors which intended to fertilise the oceans to try to obtain marketable carbon credits, the parties to the London Convention (1972) and London Protocol (1996) – which govern dumping in the high seas – developed two regulatory systems for its parties.[6] The first is a non-binding process under which the states' national environmental regulatory agency review and, if appropriate, approve an ocean fertilisation field test if it is legitimate scientific research, has undergone adequate environmental impact assessment, and satisfies other procedural requirements (London Convention and London Protocol, 2010).[7] The second is an amendment – approved but not yet in force – to the

London Protocol. Under this, its parties could either prohibit or regulate various forms of "marine geoengineering". To date, only ocean fertilisation has been so categorised by the parties, in its case as a regulated activity (London Convention and London Protocol, 2013).

In addition to treaties, countries regularly approve statements that are not intended to be legally binding but, like the statements of the CBD's parties, indicate a sense of international community. One of particular relevance is the Provisions for Co-operation between States in Weather Modification, approved by the UN Environmental Programme in 1980 (UNEP, 1980). Despite the name, its relevant definition clearly includes SRM. It is supportive of weather modification "dedicated to the benefit of mankind [sic] and the environment" (para 1.(a)), asks states to not use it to cause harm to the environment of other states and areas beyond national jurisdiction, and calls for cooperation and communication among states.

The final source of traditional international environmental law is its general principles. These remain weakly defined and not legally binding until they are operationalised in a particular agreement. For the case of climate engineering, the most relevant principles (among those that are not yet embodied as customary international law) are those of sustainable development (states should develop their resources in a sustainable manner), polluter pays (the polluter rather than the victim should pay for environmental harm and its prevention), common but differentiated responsibilities (all countries have responsibilities to prevent environmental harm but these responsibilities differ, largely based on a state's stage of economic development), and precaution (when confronting a risk of serious or irreversible harm, scientific uncertainty should not be used as a reason to postpone precautionary measures). Reasonable arguments could be made that the research or implementation of climate engineering is supported by or is contrary to each of these (Reynolds, 2014a).[8] This should not be surprising, considering the principles' inchoate character and the peculiar challenges that climate engineering presents for international environmental law, described above.

Other international law

A handful of international legal instruments outside the environmental domain warrant brief reference. Numerous observers have asserted that disagreements regarding SRM could heighten tensions among states. The UN Charter requires international disputes to be settled peacefully. Of course, if there were actually full compliance with this, then interstate hostilities would cease. Disputes are primarily political matters that may be settled through a variety of means such as negotiation, mediation, arbitration, and, in some cases, international legal forums. The legal forum with the broadest mandate is the UN General Assembly, which can take up almost any matter but issue only non-binding statements. In contrast, the UN Security Council is limited to the "maintenance of international peace and security" and can issue binding, non-consensual (i.e. majoritarian) resolutions, although five of the most powerful countries have veto power. These resolutions can be backed by the threat of force, including sanctions and military action, which would then need to be carried out by willing UN member states. The International Court of Justice is another forum for dispute resolution. Although its rulings may be enforced by the Security Council, states must consent to the court's jurisdiction in the case at hand before the trial of a contentious issue for its later ruling to be binding. Finally, some treaties contain dispute resolution forums that are applicable within their scope.

Human rights agreements provide an exception to the rule that international law governs actions that may impact other states. Under these, parties agree to treat their own citizens and residents in a manner consistent with various norms. Human rights could influence climate engineering in diverse ways. For example, states are to protect scientific freedom and to help people enjoy the benefits thereof. Climate engineering field research, or the withholding of it in the face of dangerous climate change, could affect the human rights to the highest attainable standard of health, to an adequate standard of living, and to be free from hunger.

The development of climate engineering could lead to patented inventions. Patents, which grant their holder the exclusive right to commercially utilise an invention, are domestically issued, while patent policy is to some degree internationally coordinated and harmonised. National governments may take two notable actions regarding patents as potentially controversial and important as those for climate engineering techniques. First, they may decide to exclude certain climate engineering methods from patentability because they would be contrary to public morality, including "to avoid serious prejudice to the environment" (TRIPS, 1994, art. 27.2). They may also choose to compel a patent holder to license the patent due to public interest considerations, such as on the grounds of national defence or public health.[9]

Finally, as described in the previous section, non-state instruments and institutions can be effective in regulating trans-boundary actors such as scientists. The contours of such global governance may be emerging in the case of climate engineering. Most notable has been the development of explicit, non-binding norms. Their sources are somewhat disparate: the Oxford Principles from a handful of British academics, the Asilomar Principles from the committee of a large meeting of climate engineering researchers and others, the report from a task force assembled by the US Bipartisan Policy Center, and a report issued by a think tank affiliated to the German Green Party (Leinen, 2011; Bipartisan Policy Center, 2012; Kössler, 2012; Rayner et al., 2013; see also Hanafi and Hamburg, 2018). There is remarkable overlap among these four sets, and there are no clear disagreements among them. Among other things, these variously call for public participation in decision-making, for open publication and independent assessment of results, and for climate engineering to be developed in a manner that benefits the collective public.

Synthesis and next steps

Some observers argue that because existing international law does not address all potential scenarios of conflict and harm from climate engineering, the solution is universal binding regulation of climate engineering through legal instruments. However, it may be beneficial to first take stock of extant law, the urgency of filling the legal gaps, and the limits of international law. In general, the UNFCCC regime establishes a framework for how CDR could contribute to the goal of stabilising GHG concentrations, and it might eventually offer a forum for addressing the governance of SRM as well (but see Reynolds, Chapter 20). ENMOD and the UNEP Provisions for Weather Modification point towards the international community's support of using large-scale interventions in weather and climatic systems for the benefit of humans and the environment. The CBD decision provide its sense of caution regarding climate engineering's potential negative environmental impacts while noting the need for further research. Further, universal duties concerning potential trans-boundary harm are well established in customary international law, and in some cases by specific agreements.[10] The areas beyond national jurisdiction and control each have agreements with sufficient participation and that detail their parties' rights and obligations.[11] Of these areas, the seas are the most likely site for climate engineering experimentation and implementation, and there are detailed agreements, including one with near-universal participation and a tribunal to resolve disputes. In fact, it is ocean fertilisation – the method that poses relatively large environmental risks and has seen the most progress in outdoor research – for which a detailed international regulatory regime is emerging. Finally, unilateral implementation of SRM by "rogue" countries could, in extreme scenarios, be tackled by the UN Security Council. Although not comprehensive, this is far from a legal vacuum.

In terms of urgency, most climate engineering proposals – especially relatively early field experiments – would affect the local environment first and foremost. That is the domain of national law, which is well developed in most states, and especially in those that are likely to carry out tests. Those proposed methods that might be effective and have regional or global impacts appear decades away from implementation. In contrast, large-scale field research is a more pressing matter. There, activities, risks, and effective precautions will be highly dynamic. Binding, detailed rules would quickly become obsolete, particularly in the international domain which moves more slowly. Finally, international law has limits, and not all potential international conflicts should be subject to specific legal rules. International politics – another important means to manage conflicts – may appear sloppy, improvised, and sometimes unjust, but it is adaptive and flexible. This may be precisely what's needed as climate engineering emerges.

At the same time, there are some gaps in the current international legal system that are relatively urgent but also resolvable. First, an international hub of scientific research could fulfil multiple beneficial functions (Ghosh, 2018). It could coordinate research and foster international collaboration, a low cost means to increase transparency and trust as well as to combat the nationalisation and fragmentation of research. An international body could also serve as an open repository of experiments' methodologies and results. And it could provide a site for the operationalisation of emerging research norms and possibly even their enforcement through both "carrots" and "sticks". Second, special approaches to intellectual property in climate engineering should be developed. There appears to be a consensus that patents on SRM technologies could be problematic, and alternative mechanisms should be considered before such patents become "facts on the ground" (Reynolds et al., 2017). Third, international institutions should resolve to what extent the various CDR methods could qualify towards countries' GHG emissions and for marketable credits. Lastly, a system of compensation for trans-boundary harm from climate engineering – particularly its field research – should be seriously considered (Reynolds, 2015; Horton et al., 2018).

Legal scholarship can also contribute to better understanding of climate engineering regulation. It is more than 20 years since the first academic article on climate engineering and international environmental law (Bodansky, 1996). This area has been further – and fruitfully – explored in numerous publications, especially during the last five to ten years. Yet national laws are more detailed and better enforced than international law, and most effects of early climate engineering projects will likely be experienced locally. Explorations of the implications of national law for climate engineering are an opportunity for work in the near future.

First published online as a working paper in 2015.

Notes

1 The Rio Declaration – arguably the most important general document in international environmental law – attempts to balance environmental and development goals under the rubric of "sustainable development" and the principle of common but differentiated responsibility. This latter principle is also seen in obligations to take action to prevent dangerous climate change in the UNFCCC.

2 Indeed, current modelling indicates that some forms of it could greatly reduce net climate risks at low cost and in a short time. The Intergovernmental Panel on Climate Change concluded that: "Models consistently suggest that SRM would generally reduce climate differences compared to a world with elevated greenhouse gas concentrations and no SRM ..." (Boucher et al., 2013, p. 575; see also Kravitz et al., 2014).

3 The US is not a party.

4 The treaty neither creates standing institutions nor calls for a regular meeting of its parties. Review conferences were held in 1984 and 1992, but in 2014 there was insufficient interest in a third. No complaints have ever been filed under it, and its Consultative Committee of Experts has never been convened.

5 The US is not a party but recognises most of it as customary international law.

6 Note that the London Protocol, presently with 48 parties, is indented to replace the London Convention, with 87 parties, although both are in force. Most industrialised and transitional countries are parties to at least one.

7 This was approved by a joint meeting of parties to both the London Convention and Protocol.

20

WHY THE UNFCCC AND CBD SHOULD REFRAIN FROM REGULATING SOLAR CLIMATE ENGINEERING

Jesse L. Reynolds

Some form of international regulation of solar climate engineering (solar radiation management, or SRM) is needed, both to manage its potential benefits and to minimise and possibly compensate for its harmful impacts. There is no shortage of proposals for this. Many scholars argue that this should be developed within those existing legal institutions that have (near-) universal participation, and often lean towards binding rules.

For example, Albert Lin (2009, p. 23) emphasises the mandate and expertise of UN Framework Convention on Climate Change (UNFCCC) institutions and states that its Conference of Parties (COP) should tackle climate engineering soon. Specifically, he envisions a protocol allowing for non-consensual (i.e. supermajoritarian) decision-making, "with a default presumption against the implementation of any geoengineering project". Similarly, Matthias Honegger and colleagues (2013) also argue that the UNFCCC is the logical site due to its legitimacy and scope, and the experience of its institutions. Instead of calling for a protocol, they note that the UNFCCC COP is already tacitly approving the adoption of non-consensual decisions and assert that coalitions of various sizes can operate effectively within the UNFCCC architecture. This would allow the development of guidance for solar climate engineering along with accompanying measures and with broad international support.

Partially in contrast, the staff of the Ecologic Institute advocates that the Convention on Biological Diversity (CBD) might better serve as the locus, rejecting the UNFCCC, in part because "it might be intrinsically difficult for the current climate regime to pursue a precautionary approach that is *restrictive* to geoengineering" (Bodle et al., 2014, p. 174; emphasis in original). Instead, they claim that the CBD has a mandate to minimise environmental damage more broadly – not simply to prevent climate change – and should implement "a prohibition of geoengineering activities as a general rule combined with exceptions under well-defined circumstances" (ibid., p. 135). They believe that a protocol to the CBD might be warranted for this purpose.

However, I believe that it would be counterproductive to pursue binding regulation of SRM in international forums with (near-) universal participation, and in those of the UNFCCC and CBD specifically, at least for the near future. In general, it is too soon to initiate global negotiations towards a binding agreement. The proposed technologies remain "imaginaries" and what they may be able to accomplish, how they would operate, their reversibility, their costs, and their risks all remain uncertain. What understanding we do currently have is from relatively simple and extreme implementation scenarios (see, for example, Kravitz et al., 2011, 2013, 2015; Irvine et al., 2016).[1] Moreover, we are far from developing political consensus regarding what we may (not) want from SRM. The lack of knowledge and agreed-upon objectives would lead to highly divergent state interests and an absence of negotiation focal points. Any resulting binding agreement developed in the near term would lock us into commitments that may later seem unwise.

Moreover, consideration of SRM regulation in an international forum with (near-) universal participation is problematic. If representatives were to be aware of and rationally consider their countries' vulnerabilities to climate change and how they might gain (or lose) through potential SRM implementation, then global negotiations might be potentially fruitful. However, this ideal is not the case now, in part due to the current low state of knowledge described above. Under this condition, the more numerous developing countries may fear permitting industrialised countries to pursue a technology that, from the former's perspective, would offer uncertain benefits while possibly giving the latter an excuse to delay mitigation and granting them the power to shape other countries' climate (but see Reynolds, 2015 as well).

Indeed, this appears to have been the case at the 2010 CBD COP, which produced a poorly worded, restrictive statement at the motivation of some developing countries.[2] This understandable predisposition against SRM is exacerbated by the pessimistic tone of the existing mainstream and academic discourses, which tend to emphasise climate engineering risks and obscure its potential to reduce climate change.[3] A prohibition, perhaps with only narrow exceptions, is a foreseeable result.

This would be undesirable because SRM does appear to hold the ability to lessen climate change risks, which are more severe in developing countries. A ban could also push field research to less responsible states, and may cause any eventual implementation – perhaps in response to sudden climate change – to be carried out based upon a comparatively thin knowledge base (Victor, 2008, p. 325; Parker, 2014). Nevertheless, the countries that are interested in pursuing SRM research and may eventually have the capacity to implement it – which also tend to be relatively powerful – would likely not concede to such a proposal, resulting in either a stalemate, language that would be vague to the point of little use, or a prohibition without the participation of the countries with implementation capacity (Victor, 2008, p. 331).

Specifically, the UNFCCC possesses some particular drawbacks. First, its negotiations are already highly politicised, and arguably dysfunctional; stirring the pot now with SRM is unlikely to be fruitful. Second, the negotiators and staff there appear strongly committed to the dominant paradigm of mitigation and adaptation, and the institutional culture might be hostile to SRM.[4] Finally, several existing and proposed provisions currently under the UNFCCC, such as the Green Climate Fund, the Loss and Damage Mechanism, technology transfer, and the Clean Development Mechanism would transfer wealth from rich to poor countries.[5] Potential recipients of these transfers may believe that SRM could undermine the justifications of these mechanisms, and consequently resist its serious consideration.

The CBD fares worse. It would be a stretch of its mandate to develop detailed regulations for activities to reduce climate risks. If the CBD were to attempt this, it would need the close cooperation of the UNFCCC, whose staff may feel that its administrative domain is being infringed upon (Bodle et al., 2014, pp. 134, 174).[6] Moreover, if the experience of genetically modified organisms is any guide, the politics of the CBD may cause its climate engineering policy to be based upon opposition to the practice itself rather than a weighing of its potential benefits for and risks to biodiversity (Strauss et al., 2009, pp. 519–520; Honegger et al., 2013, p. 129). Finally, the Unites States – the world's leading research state – is not a party to the CBD.

It is better to conceptualise the unfolding of international regulation of SRM as a process instead of a singular, final, and known destination. If the technologies are actually developed, they will pass through various stages, each presenting different problem structures (Reynolds, 2014a, pp. 284–288). In the short term, we need more knowledge of their capabilities, risks, means, costs, and reversibilities. This can be improved through research, including field tests which can gradually and cautiously increase in complexity, scale, and perturbation. For the time being, the risks of these can be managed through existing national and international environmental law, institutions, and norms.[7] This research should be internationally coordinated, but not made monolithic in a manner that drowns out sceptical views.

Meanwhile, we need to work towards consensus as to what we do and do not want from climate engineering and its research. This requires engagement with the public and policymakers: for well-informed, balanced debate, and for the continuation of norm development. At some time, these norms should be operationalised into more detailed guidelines and best practices. In these processes, an international institution could help to facilitate and coordinate research, to foster international cooperation, to provide a site for norm development and operationalisation, and to help ensure that field experiments are responsibly conducted. This need not be highly legalised, but if field work increases in scale and perturbation, greater legalisation would be warranted.[8] Regardless, all the researching countries should be represented here.

Looking much further ahead, if a deployable SRM technology were eventually ready to be used, an international institution that takes a managerial approach, described above, might be adequate to prevent its misuse, but a multilateral agreement may ultimately be warranted. Even if only a few countries would be capable of global implementation, and would thus be the only ones which *must* participate in its regulation, for both normative and political reasons a larger – although not necessarily universal – forum would be preferable. The UNFCCC institutions, or perhaps those of the CBD, may or may not turn out to be an effective site for this.

Regardless, I assert that it is presently not a productive endeavour to dwell on how states might collectively govern technologies which do not yet fully exist; whose forms, benefits, risks, costs, and reversibilities remain unknown; and under what circumstances and for what purposes they might be used still uncertain. Indeed, this focus can even be counterproductive, if fears of an intractable, distant, arguably unlikely future hinder a less problematic present course which may lead to the reduction of human suffering and environmental degradation.

First published online as an opinion article in 2015.

Notes

1 The scenarios of the largest modelling project to date, the Geoengineering Model Intercomparison Project (GeoMIP) keep radiative forcing or the magnitude of SRM constant. The scenarios either stop abruptly or continue indefinitely. They do not balance residual temperature and precipitation anomalies, they vary with the time of year or with latitude, they merely slow down the effects of climate change, and they do not phase out gradually. Many of them use an impossibly great atmospheric carbon dioxide concentration in order to produce clearer results. This is not intended as a critique of the project, but instead to highlight the limited state of current modelling.

2 Sugiyama and Sugiyama, 2010, p. 8 note that "the information at the [COP] site was very limited and delegates were not well informed about the science of geoengineering". ETC Group, 2018 report that "The push ... came largely from governments of the global South".

3 This perception is empirically supported. For example, among policy documents (including national and international, as well as governmental, intergovernmental, and non-governmental) which discuss climate engineering, more than twice as many express concerns than they do hopes Huttunen et al., 2014, p. 10). An unpublished survey of 101 articles on climate engineering in 18 major news sources in three countries indicates that, among articles which discuss SRM, 50 per cent are negative, 47 per cent are balanced, and 3 per cent are positive. Twenty-three per cent of these SRM articles mention no potential benefits while 94 per cent bring up one or more risks (Elblaus, 2014, pp. 4, 7, and personal communication). David Keith found that roughly half of all results from a Google search on climate engineering discussed the concern that SRM would dramatically reduce the summer monsoon in south Asia, a result from early modelling which more recent work has shown to likely be a less severe problem (Keith, 2013, p. 55). In the academic discourse, I am most familiar with international law, where almost all reviews consider how climate engineering should be regulated to reduce its own risks yet do so without considering how it might reduce climate change risks (Reynolds, 2014a, pp. 427–434).

4 On the other hand, the UNFCCC has become increasingly receptive to adaptation, which for many years was somewhat off-limits. The situation with regard to SRM could similarly change.

5 For a discussion of these transfers' rough magnitudes and some problematic implications, see Posner and Weisbach, 2010. This is a key criticism of emissions permits among economists. (See Cooper, 2010.)

6 Although Bodle et al. discuss regime complexes, in which multiple international institutions govern a particular issue area, they provide no specifics as to how the CBD and the UNFCCC would cooperate.

7 Norms and other forms of soft law are important in international regulation, particularly for technical and dynamic phenomena and for those undertaken by transnational actors, both of which are characteristics of climate engineering research. These norms are emerging for climate engineering, as seen in Bipartisan Policy Center's Task Force on Climate Remediation, 2012; Leinen, 2011; Rayner et al., 2013. As another example, human subjects research is internationally governed by non-binding, non-state documents such as the Declaration of Helsinki.

8 See the examples of internationally coordinated scientific research in Ghosh (Chapter 31 in this volume). An example of an international institution with a managerial approach to scientific research and the responsible conduct thereof is the International Atomic Energy Agency. (See Reynolds, 2014b.)

21

SOLAR GEOENGINEERING AND THE PROBLEM OF LIABILITY

Joshua B. Horton, Andy Parker, and David Keith

The prospect of solar geoengineering, or methods of reducing incoming solar radiation in order to offset the effects of climate change, raises a number of significant governance challenges that are likely to be far more difficult to overcome than the technical barriers to development. Who decides when to use solar geoengineering? Who decides when to test it? How should these decisions be made? What climatic targets should guide an intervention? How should geoengineering be tied to mitigation and adaptation? What if something goes wrong?

This last question points to the issue of liability, that is, how society resolves cases in which parties suffer unfairly from the consequences of activities carried out by others. If solar geoengineering were deployed, it would yield both benefits and harms because, even if it could perfectly compensate for greenhouse gas driven climate change (which it could not) it would harm those who stand to gain from a warming world. Such harms would be expected in that they would arise from the intended effect of geoengineering. Unintended damages might occur if geoengineering acted unexpectedly. A just and stable governance regime will require some way to compensate countries for damages incurred.

A system of liability and compensation would need to be based on sovereign states since even if private actors undertook geoengineering, national governments would likely be held ultimately accountable for their actions. Undoubtedly, it will be more difficult to get widespread agreement to conduct geoengineering without a liability mechanism.

Take the following hypothetical situation. Twenty years from now, the United Nations (UN) decides to authorise and endorse a large-scale stratospheric aerosol injection programme headed by the United States (US). The programme is intended to create a thin aerosol "sunshade" in the upper atmosphere that will block a small fraction of sunlight from reaching the Earth's surface, thereby reducing global temperatures to counteract increased greenhouse gas concentrations. Within months of deployment, an unusually severe drought strikes Russia, its annual grain harvest falls far below expectations, and millions of Russians are threatened with hunger and malnutrition. The Russian government blames these events on the UN-sponsored, US-led solar geoengineering programme initiated months earlier, and demands restitution. How would the international community deal with this?

Many scenarios like this can easily be imagined. In this case, the available scientific evidence may not allow for definitive answers to the question of what caused such extreme weather. Despite weak evidence, the Russian government may feel pressured by its populace to hold geoengineering responsible. More cynically, the government may suspect – or know – that there is no causal linkage between geoengineering and the drought in Russia, but may publicly blame the former anyway in an attempt to extract international reparations. Or Russian scientists may sincerely believe that a strong causal connection exists, leading the government to make the charge in good faith.

The real possibility of such situations arising means that any workable solar geoengineering governance regime will need to include some arrangement for assessing liability for alleged damages and awarding compensation when appropriate. Furthermore, major world powers would be unlikely to accept a geoengineering deployment in the absence of a well-designed liability system. Therefore, a credible liability and compensation mechanism will be essential to any future use of solar geoengineering techniques, and arguably to any future field research with the potential for substantial cross-border effects.

Precedents for international liability

Researchers have identified many possible effects of solar geoengineering that could pose risks across international borders. Disruptions to regional hydrological cycles, like droughts, are one widely discussed risk. Others include damage to the ozone layer, increased acid rain, negative effects on plant ecology, and reduced effectiveness of solar power generation (Royal Society, 2009). If implementation is ever seriously considered, these and other risks associated with stratospheric aerosols, marine cloud brightening, and possibly other forms of solar geoengineering will need to be adequately managed via a system of international legal liability.

Fortunately, the international community has considerable experience of crafting and executing liability regimes, and legal scholars have developed an impressive body of knowledge about the theory and practice of international liability. In international law, liability systems vary in two key ways. First, the standard of liability, or principle by which culpability is assigned, can be either "fault-based", according to which a country must have caused cross-border damages *and* must have done so intentionally, or "strict", for which cause alone is sufficient to establish liability. Second, accountability for damages can be structured as either "civil", under which private parties are held responsible, or "state", in which case governments are responsible. Liability regimes also differ in terms of exemptions allowed, the extent of damages third parties may claim, the extent of compensation responsible parties must provide, and in other ways.

In practice, most liability regimes have been based on strict, civil liability, and the majority of regimes have been developed in the environmental field. Starting in the 1950s, international liability regimes have been adopted to address oil spills, movements of hazardous wastes, nuclear accidents, protection of the Antarctic environment, genetically modified organisms (GMOs), and other environmental hazards, as well as non-environmental issues such as aircraft accidents and space debris (Secretariat of the International Law Commission, 1995). The histories of some of these regimes provide useful insights into the sorts of institutional features that would be appropriate to a future system of liability for solar geoengineering.

The Space Liability Convention of 1972, for instance, set up a system for compensating third parties who suffer damage from spacecraft, satellites, or other "space objects", whether the damage occurs in outer space, in the atmosphere, or on the ground. The Convention carves out a special role for "launching states" as potentially liable parties, a designation with obvious application to solar geoengineering activities. In 1981, the claims procedure prescribed by the Convention helped lead to an amicable settlement between Canada and the Soviet Union after the Soviet nuclear-powered satellite Cosmos 954 accidentally broke up over Canadian territory.

Another useful example is the international oil spill regime, which consists of a series of interlocking agreements adopted, beginning in 1969. The core functions of this arrangement are provided by the International Oil Pollution Compensation (IOPC) Funds, a collection of independent financial bodies responsible for providing indemnification to parties who suffer damage as a result of oil spills at sea. The IOPC Funds are governed multilaterally but funded exclusively by mandatory levies on oil companies; to date, the funds have paid out more than $700 million in compensation (IOPC Funds, 2011). Such a financing mechanism, built explicitly on the "polluter pays" principle, might serve as a model for solar geoengineering liability since climate interventions like stratospheric aerosols would be staged in response to carbon pollution largely traceable to the fossil fuel industry.

The problem of attribution

When considering liability in the context of solar geoengineering, an especially problematic issue has to do with the difficulty of demonstrating causal attribution. Traditionally, proving legal liability has required showing a direct cause-effect relationship between an action committed by an alleged wrongdoer, and damages suffered by the victim. Determining the source of an oil spill is a relatively straightforward task. However, establishing an unambiguous direct causal connection between a solar geoengineering intervention and damages alleged to have occurred as a result is impossible due to the highly complex nature of the climate system. Instead, interventions in the climate system (*including* greenhouse gas emissions) affect the *likelihood* of particular weather events or other outcomes occurring, by shifting the systemic parameters within which discrete events take place.

The inapplicability of traditional notions of attribution in the climate field presents obvious obstacles to constructing a workable system of liability and compensation with regard to both climate engineering and climate change more broadly. Yet this problem has been successfully overcome in other fields such as tobacco litigation, "toxic torts" (including pharmaceuticals), and radiation exposure, where statistical evidence based on probabilistic models has served as the basis for findings of liability and damage awards. Probabilistic approaches such as Fraction Attributable Risk (FAR) are currently being developed and refined by climate researchers, and are increasingly being used to assert causation and liability in the growing body of climate law (Lord et al., 2013). In the US, for instance, claims of climate damage based on statistical models have featured in high-profile cases including *American Electric Power Company v. Connecticut, Kivalina v. ExxonMobil Corporation,* and *Comer v. Murphy Oil.* In the UN Framework Convention on Climate Change (UNFCCC), attribution is an important focus of the ongoing Work Programme on Loss and Damage (Approaches to address loss, 2014).

Design principles

A careful consideration of existing liability regimes and evolving approaches to attribution using statistical methods points towards a small set of first-order design principles for any future solar geoengineering liability regime:

1. Strict liability – This has become the accepted standard for international liability.
2. State accountability – Governments, rather than private parties, would be primarily responsible for staging an intervention.
3. Compensation fund – Similar to the IOPC Funds, potentially financed by the fossil fuel industry (because the underlying climate risk for which geoengineering is a partial and imperfect fix derives from fossil fuel emissions).
4. Attribution based on probabilistic models – FAR (or preferably more sophisticated methods) would be needed to demonstrate causation.
5. Inclusiveness – A wide range of governments, businesses, and other actors involved with the regime would help ensure adequate resources as well as political legitimacy.
6. Flexibility – Institutional flexibility would be key to coping with the multiple uncertainties associated with climate science and geoengineering, in addition to geopolitical unknowns.

These principles, while high-level, are informed by historical practice, established jurisprudence, and current scientific understanding, and would surely

PART V

National, regional, and sectoral perspectives

Introduction

Several chapters in this book have highlighted the potential for geoengineering technologies to significantly alter global politics. However, just as the concept of geoengineering is not emerging in an institutional vacuum, neither is it emerging in a political one. Existing political perspectives on issues ranging from climate change to economic development are likely to shape the way geoengineering proposals are considered and discussed in national and regional contexts. It is entirely possible that, rather than geoengineering remaking global politics, existing geopolitical narratives end up shaping the development and potential use of geoengineering technologies.

To explore this challenging topic, this section of the book is comprised of seven contributions reflecting diverse global perspectives on geoengineering. The first four reflect concerns and dynamics particular to countries or geographic regions; the latter three are organisational or sector-based perspectives. A standard caveat is necessary: the authors do not, of course, presume to speak as representatives of the wider demographics or organisations in which they live and work. Nevertheless, they offer embedded and educated perspectives that give us the opportunity to consider political perspectives that could shape the evolution of geoengineering debates and technologies, as well as the importance of forums in which such perspectives can be widely shared and discussed.[1]

In Chapter 22, Mulugeta Ayalew and Florent Gasc kick off with their perspective on the potential implications of geoengineering for Africa. They provide a nuanced analysis of the possible benefits and risks for African communities while emphasising the need for African countries to have a prominent voice in future decisions about the evolution and use of geoengineering technologies. In Chapter 23, Weili Weng and Ying Chen challenge previous speculations that China "would unilaterally resort to" geoengineering. Building from their analysis of China's longstanding weather modification activities, they call for cautious research to reduce uncertainties built upon international capacity-building "to eliminate misunderstanding and promote cooperation".[2] In Chapter 24, Penehuro Lefale and Cheryl Anderson reflect on the "political, policy, and scientific dilemma" that geoengineering poses for small island states. Reflecting on conversations about geoengineering by representatives of Small Island Developing States (SIDS) in the Pacific, they highlight the need for inclusive international governance mechanisms to avoid "side-lining small island issues" in the global geoengineering discourse. And in Chapter 25, Simon Nicholson and Michael Thompson turn their attention to the US context, reflecting on how geoengineering technologies could turn on its head the existing left–right (Democrat-Republican) divide on climate policy, and examining how the American political landscape might influence the evolution of geoengineering research.

In Chapter 26, rather than focusing on a geographically defined population, Pablo Suarez and colleagues examine geoengineering from the perspective of international humanitarian organisations. Starting from the question of "What role, voice or agency will the vulnerable have in geoengineering decisions?" Suarez et al. articulately advocate that future geoengineering research and deployment be managed in ways that care for those most vulnerable in our global society. In Chapter 27, once again moving beyond geographical borders, the Action Group on Erosion, Technology, and Concentration articulates the perspective of the "Hands Off Mother Earth" campaign that the "uncertainties and inadequacies" surrounding proposed geoengineering technologies are so large that they should not be developed for deployment.[3] Finally, in Chapter 28, Chad Briggs explores solar radiation management (SRM) geoengineering from the perspective of national security planners and argues that SRM is not likely for now to become weaponised as an option for altering environmental conditions, as it is neither tightly controllable nor a tool of proportional response. Nevertheless, its prospective deployment for non-military purposes remains a background concern for more traditional security operations.

This section is, of course, illuminating for what and whose perspectives are *not* covered, as much as for those that are. Collectively, these seven contributions represent only a tiny fraction of the diverse perspectives relevant to the global geoengineering discourse, and at a particular moment in time. As geoengineering approaches mainstream political discussions, we must be watchful for emergent concerns and constituencies in politics and society (this section) and in legal frameworks (see Part IV) and continue to ask ourselves how these perspectives reflect the public, stakeholders, and framings at play (Part III).

Notes

1 The Solar Radiation Management Governance Initiative has held workshops in numerous developing countries and emerging economies to more fully incorporate

their concerns and perspectives into the debate. A list of their engagements – many with summaries of key points – can be found at: http://www.srmgi.org/events/.

2 A Chinese geoengineering research programme has since been initiated with the support of the National Key Basic Research Program, examining processes and impacts, risk and governance of geoengineering approaches (Cao et al., 2015; Moore et al., 2016). It is worth noting that its structure and research scope does not differ from antecedent and current research programmes based in North America or Europe.

3 It is worth noting that the ETC Group's frequently repeated claim – that the 2010 decision of the Convention on Biological Diversity constitutes a de-facto moratorium on geoengineering research – is a highly partisan interpretation (see Reynolds, Chapter 19, for a more legalistic reading).

22

MANAGING CLIMATE RISKS IN AFRICA

The role of geoengineering

Mulugeta Mengist Ayalew and Florent Gasc

A given community could approach geoengineering in one of three simplified ways. The first is to work for an international ban of geoengineering research, development, and deployment. Second, it may advocate for a laissez faire approach to relevant activities. Third, it may work towards an international system of governance and regulation. Africa should not only opt for the third option but also actively work in shaping the form and content of such international regulatory system as it may emerge.

Climate change adversely and particularly affects Africa, which has limited adaptive capacity. By reducing the length of the growing season and putting marginal lands in large areas out of production, climate change is projected by 2020 to reduce yield by as much as 50 per cent in some countries, and to potentially expose hundreds of millions of additional people to risks of water stress. Coastal population centres will be adversely affected by an expected sea level rise (Niang et al., 2014).

The fact is: the structure of the problem is such that the fate of Africa is largely in the hands of the big polluters. Discussion of climate risks should recognise Africa's "helplessness"; much of the problem is not Africa's doing and hence there is nothing much it can do to reverse the situation, apart from engaging in adaptation planning. Adaptation activities, as response to climate risks in Africa, are fraught with problems of resources and limited effectiveness. It is therefore important not to downplay Africa's role on the mitigation front. One thing it cannot afford to do is to play into the "tragedy of commons" characterising current global mitigation efforts. Working to realise available mitigation potential could help Africa occupy a higher moral ground and hopefully will put pressure on the big polluters to take action.

It is in this context that the implications of geoengineering should be seen. By claiming to limit and even reduce temperature, geoengineering will address the main cause of Africa's vulnerability. Adverse effects of climate change are caused by not only temperature changes but also changes in precipitation and climate extremes. Hegerl and Solomon (2009) critique the current debate on geoengineering for its focus on limiting warming. While it is true that the desirability of a given course of action should be evaluated on its effectiveness in addressing changes in all aspects of the climate system, one must not also lose sight of the relative weight of temperature rise as compared to changes in precipitation patterns and frequency and severity of climate extremes. It is this latter consideration which provides a *prima facie* case for consideration of geoengineering for its use to buy time to address the root causes of the problem (Wigley, 2006) and to manage risks more severe than anticipated.

In Africa, relative to changes in precipitation pattern and extreme events, it is the temperature rise that is expected to cause extreme human suffering. Food security, health, and biological diversity will be adversely affected. The issue of food security is given such importance in the current climate change discourse that it is expressed in the ultimate objective of the United Nations Framework Convention on Climate Change. Studies suggest that it is temperature rise (as opposed to changes in rainfall patterns) which will largely impact food production by adversely affecting pollination, grainfilling, and photosynthesis (Araus et al., 2008; Burke et al., 2009; Schlenker and Lobell, 2010).

Geoengineering not only claims to address the principal concern of Africa, temperature increase, but it also appears that it might reinstate Africa's fate largely into its own hands. Possibilities of unilateral deployment of readily available and cheap technologies are what make solar geoengineering through stratospheric aerosol injection seductive, compared to related technologies and mitigation efforts. It is these aspects also which could potentially restructure the problem, giving Africa a meaningful role in managing climate risks on its own. We do not have any illusion that even the lower costs and readily available technologies of geoengineering are within easy reach of any of the African countries. Any obstacles in this regard can, however, be overcome with appropriate cooperative arrangements among countries in Africa. Any deployment of geoengineering by African countries could also be undone by counter deployment of such or similar technologies by one or more countries. Despite this possibility, however, the negotiating position of Africa and hence control of its destiny could be enhanced.

There, therefore, lies the *prima facie* attractiveness of geoengineering in minimising climate risks and development challenges in Africa. This should not, however, prevent consideration of geoengineering risks. The principal claim of geoengineering is that, by reducing incoming and/or increasing outgoing short-wave radiation, it will cool the Earth. The evidence for this so far is based on effects of volcanic eruptions, sulphate pollution, and modelling exercises. It might therefore be prudent to see any harmful side effects of geoengineering by studying effects of volcanic eruptions, for instance. Conceptually, it is plausible to expect that reducing energy absorbed by the earth will affect the level of evaporation, further reducing precipitation. After examining the precipitation and stream flow records from 1950 to 2004 and taking into account changes from El Nino, Trenberth and Dai (2007) concluded that the eruption of Mount Pinatubo in June 1991 resulted in "substantial decrease in precipitation over land and a record decrease in runoff and river discharge into the ocean from October 1991– September 1992 [sic]". The authors also found that the changes are greater in the tropics, which has worrisome implications for Africa. Other studies also found or predicted that stratospheric aerosols might expose regions including Africa to drought conditions (Narisma et al., 2007; Robock et al., 2008). Climate change is expected to expose millions of Africans to risks of water stress. It is not clear whether any drought resulting from geoengineering would be additional to the changes in rainfall patterns expected from climate change. If not, careful trade-offs have to be made in this regard, a decision which will benefit from more refined scientific knowledge. Other risks mentioned include ocean acidification, ozone depletion, moral hazard, and termination (MacCracken, 2006; Matthews and Caldeira, 2007; Robock, 2008; Robock et al., 2008).

From the discussion above, Africa's interest lies not in a laissez-faire development and deployment of geoengineering. Nor does it lie in a ban on geoengineering. It will be recklessness of the highest magnitude to tie the fate of Africa to mitigation efforts by developed and major developing countries. The challenges of taking this course are blindingly obvious. Because of collective action problems, the reduction in emissions of greenhouse gases required to keep the increase in temperature rise less than 2°C below pre-industrial levels has proved elusive, at least so far. Even though some unlikely breakthrough is found and such reductions are made, this will still expose human lives in Africa to unacceptable risks. It therefore benefits Africa, perhaps more than any region of the world, if efforts are expended to develop options to contain temperature rise should global mitigation efforts prove inadequate or should unaccepted climate risks materialise.

Any governance and regulatory system that seeks to account for Africa will have the following elements, among others. First, it should principally work towards a progressively better understanding of the effects of geoengineering in minimising and maximising climate risks in Africa. Any regulatory regime addressing a novel problem (such as the initial climate change convention) is bound to be a system of information gathering and reporting, at least in the beginning. A number of instruments could be developed to achieve this purpose: disclosure obligations on those engaged in relevant activities and financial support for studies of geoengineering risks in Africa, just to mention two. Second, such a regulatory system should work towards the development of "safe and effective technologies", that is, adverse effects of the technology on the continent should be minimised to a possible minimum. There are suggestions that possible risks of geoengineering such as risks of termination, human health risks of sulphur deposition, and destruction of the ozone layer could be avoided or minimised through better engineering. Third, it should identify financial resources for geoengineering research from sources that do not reduce the flow of climate finance to Africa. Fourth, it should recognise the right of Africa to have access to safe technologies. Fifth, it ought to design ways of compensating Africa should any deployment (sanctioned or otherwise) results in loss and damage. Sixth, it should recognise the paramount role of mitigation as the appropriate response to climate change and should circumscribe the role of geoengineering for managing emergencies in cases where the climate changes faster than expected and where global mitigation efforts prove inadequate.

No single course of action is without uncertainties and risks. The question is how each fare with its benefits and risks compared to others. Such assessments require extensive knowledge and information as to the performance of alternatives. Africa's interest, considering potential roles of geoengineering in minimising and maximising climate risks on the continent, requires generation of extensive scientific knowledge and control of its deployment with a view to minimising costs of errors.

First published online as an opinion article in 2013.

23

A CHINESE PERSPECTIVE ON SOLAR GEOENGINEERING

Weili Weng and Ying Chen

With emissions of greenhouse gases (GHGs) continuing to escalate, scientists and policymakers are giving increasing attention to solar geoengineering or solar radiation management (SRM): a suite of hypothetical technologies with the potential to dramatically alter the dynamics of climate change and cool the climate. The Summary for Policymakers of Working Group 1 in the IPCC's Fifth Assessment Report states: "modelling indicates that SRM methods, if realizable, have the potential to substantially offset a global temperature rise" (IPCC, 2013). But the report also indicates that "the level of understanding about [reducing incoming solar radiation] is low, and it is difficult to assess feasibility and efficacy because of remaining uncertainties in important climate processes and the interactions among those processes."

The mention of geoengineering in this report has been interpreted as recognition of theoretical "Plan B". Some believe that it reflects growing governmental interest in the capacity of these ideas to address climate change if governments fail to adequately reduce GHG emissions. Concerns are centred on governments which would unilaterally experiment with large-scale Earth system manipulations, purportedly to cope with climate change.

It has often been speculated that China, as one of the world's largest and fastest growing GHG emitters and a significant player in climate negotiations, would unilaterally resort to this "Plan B". There are some debates on why and how China would carry out geoengineering, but none of these arguments have been supported by solid evidence. Such purported rationales are extrapolated from China's social, political, and cultural aspects, and undue judgements are made on the possibility of its actions on geoengineering. This article will examine these assumptions and try to explain that with currently available information, these arguments on China's position and strategies towards geoengineering can be fallible.

As one of the major emerging developing countries, China plays an important role in the international climate negotiations and faces a very serious challenge in emission abatement. According to the International Energy Agency (2012), China emitted 7.26 billion tons of CO_2 from fossil fuel combustion in 2010, accounting for about 23.8 per cent of the world total, while the United States accounted for 17.7 per cent, and the EU accounted for 12.1 per cent. Supposing that China maintains a rapid economic growth of 8 per cent per year, it would probably emit nearly ten billion tons of CO_2 in 2020, even if the target of a 40–45 per cent decrease in carbon intensity has been realised. Hence, the Chinese government has always attached great importance to addressing climate change issues.

Although China's efforts in CO_2 mitigation have been recognised as substantial, there are concerns that these efforts – in addition to the efforts of other states – may not be enough to slow the warming of the globe. Some, therefore, surmise that China might be tempted to carry out geoengineering activities. There are media comments that list China as one of the four countries – "the US, Russia, China and Israel – who possess the technology and organization to regularly alter weather and geologic events for various military and black operations" (Andersson, 2012). For example, Clive Hamilton, a professor of public ethics at Australia's Charles Sturt University and a prominent critic of geoengineering, suggests in his latest book *Earthmasters: The Dawn of the Age of Climate Engineering* that China might be one of the most likely candidates to go it alone with an SRM technique like sulphate aerosol spraying (Boyd, 2012, referring to Hamilton, 2013a). The reasoning process goes: "China is highly vulnerable to water shortages in the north, with declining crop yields and food price rises expected, and storms and flooding in the east and south. Climate-related disasters in China are already a major source of social unrest so there is a well-founded fear in Beijing that the impacts of climate change in the provinces could topple the government in the capital", and therefore, "the political dilemma over geoengineering will perhaps be most acute in China" (Hamilton, 2013b).

It is correct that China is vulnerable to negative climate impacts, and is under tremendous pressure to reduce GHG emissions while maintaining rapid economic growth. However, there is no direct causal relationship between the fear of social disruption and the resort to geoengineering. Taking into consideration the immense uncertainty associated with SRM techniques, it would be more logical to exclude geoengineering in the options to appease social unrest.

Besides the fear of social disruption, another typical emphasis on Chinese temptations towards geoengineering is that China has been paying a lot of attention to the use of artificial weather technologies, particularly in drought and water management. Therefore, China has the capacity and disposition to unilaterally carry out geoengineering.

It may be the case that China has done some weather modification programmes. But as revealed by Fleming, technological interventions into local weather have a long and chequered history and are employed in dozens of countries in some form or other (Fleming, 2010). Practices to stimulate artificial rainfalls are occasionally reported in the Chinese media, especially in order to ensure good weather for large-scale public events such as the 2008 Olympics in Beijing, and thus were extensively exposed to the international media. As a result, some critics and even meteorological scientists may conflate these weather modification techniques with a generalised notion of "geoengineering", and raise various questions over unilateral implementation.

Compared with some other countries that confine weather modification technologies to the experimental stage, China uses weather modification technologies relatively frequently. However, weather modification technologies in China are strictly implemented only under specific weather conditions, and in a small area range. These activities are no more than an accelerated process of precipitation. Only when there are mature precipitation conditions and strong necessity will there be a limited scale and time period of artificial intervention. Moreover, all these activities are subject to the Regulation of Weather Modification, issued by the Central People's Government of China (CPG, 2005). It stipulates that weather modification means can only be used to avoid or mitigate meteorological disasters, such as drought, under appropriate conditions. China cannot carry out unilateral climate engineering simply because it attempts to control the weather on a local and seasonal scale. There are levels of magnitude in between local weather control and global geoengineering in terms of scientific knowledge, and human, financial, and material resources.

It was only at the end of 2012 that China began to list geoengineering in its supporting category of the National Natural Science Foundation of China (NSFC, 2012). This decision can be interpreted as a desire to develop a national capacity for understanding or keeping up with research in the Western countries. In fact, there are very few research articles on geoengineering-related topics in Chinese academia. These articles have mostly emerged in the last handful of years; many of them are observations about research and debates in the Western countries and have expressed a good deal of scepticism about geoengineering.

Currently, China has focused more on carbon capture and storage technologies (CCS), though these are still in the "introduction-trial" demonstration stage and far from being commercialised. This may demonstrate a potential interest in particular forms of carbon dioxide removal (CDR), but not in SRM. Indeed, in contrast with many studies conducted in Europe and North America, China has not yet commenced modelling or field tests into any suite of geoengineering technology. Top climate change scientists have all expressed concerns over the risks of geoengineering. Ding Yihui, the special adviser on climate change of the China Meteorological Administration (CMA), has noted: "If human beings continue to emit GHGs, the implementation of geoengineering could only change short-wave radiation, while the Earth's long-wave radiation problem which is caused by carbon dioxide and other greenhouse gases remains not solved" (Ji and Zhong, 2010). An article in the *Pacific Review* on China's "blunt temptations of geoengineering" also expressed a similar opinion: "While Chinese climate scientists are keenly aware of the potential benefits of geoengineering as well as risks, there is no significant constituency currently promoting unilateral implementation of SRM" (Edney and Symons, 2013).

There is a gap in the research on the science and regulation of geoengineering between China and most research activities that have actually taken place in developed countries. Lack of adequate dialogues with the international academic community and media may contribute to the speculations on China's perspectives. As geoengineering has global implications, developing countries like China need to be included in international discussions about research and governance activities. An initiative begun in 2015 – a coordinated geoengineering research project funded by the National Key Basic Research Program of China on physical mechanisms, climate impacts, and risks and governance – may help incorporate Chinese expertise and perspectives into increasingly global conversations (Cao et al., 2015). Good governance mechanisms are indispensable in order to make sure that any research that proceeds is safe, transparent, inclusive, and responsible, and to encourage international cooperation. China should seek support for capacity-building in this area, and actively engage in the development of governance guidelines or rules under existing international treaties, institutions, or regimes, which are still rudimentary at the current stage, to eliminate misunderstanding and promote cooperation.

First published online as an opinion article in 2014.

24

CLIMATE ENGINEERING AND SMALL ISLAND STATES

Panacea or catastrophe?

Penehuro Fatu Lefale and Cheryl Lea Anderson

Climate engineering research may soon no longer be confined to the domain of a few nations. It may even be only a matter of time before the topic emerges into international climate, trade, and development policymaking. Recent years have seen dramatic increases in media coverage and publications on the topic (Beiter and Seidal, 2013). Policy dialogue focuses on the conceptualisation or the modelling stages of climate engineering, but engagement in climate engineering field experiments and demonstrations also demands urgent attention. Increasing evidence now suggests there is a growing need for the development of an international governance mechanism to regulate the research and deployment of climate engineering techniques with complex and widespread impacts – perhaps a protocol(s) or other legal instrument(s) on climate engineering under existing international treaties and conventions (e.g. UN Framework Convention on Climate Change or the Convention on Biological Diversity, as at the London Convention and Protocol in the case of marine geoengineering).

For small islands, the emergence of climate engineering as a possible supplement to mitigation and adaptation strategies presents a political, policy, and scientific dilemma. They must engage, without delay, in the international climate engineering conversation. Climate engineering – in particular, solar radiation management (SRM) – will likely have profound effects on them. It could be the panacea they are searching for in their quest to stabilise the climate or a potential anthropogenic catastrophe in the making.

Climate change is the most serious challenge facing small islands. Greenhouse gas emissions from small islands are negligible in terms of global emissions but the threats of climate change and sea-level rise to small islands are very real (Mimura et al., 2007; Nurse et al., 2014). Indeed, the very existence of some atoll nations is threatened by rising sea levels even though these island nations did not contribute to the problem. Furthermore, impacts of climate change on small islands will have serious negative effects on socio-economic and bio-physical resources – although some local impacts may be reversed through effective adaptation measures (Nurse et al., 2014).

In addition to climate change, there is increasing recognition of the risks and negative impacts on small islands from climate-related processes originating well beyond the borders of an individual nation or island. Such trans-boundary processes include airborne disease from the Sahara and Asia, distant-source ocean swells from mid-latitudes, increases in plant and animal species and the spread of aquatic pathogens.

Small islands are disproportionately affected by current hydro-meteorological extreme events, both in terms of the percentage of the population affected and losses as a percentage of GDP (Mimura et al., 2007; Nurse et al., 2014). Under climate change, the risk of damage and associated losses is expected to continue to rise. Much of the existing literature on climate risks in small islands focuses on managing present-day risks, rather than future high risks, through risk transfer, risk spreading, or risk avoidance (Nurse et al., 2014). Risk transfer is largely undertaken through insurance, risk spreading through access to and use of common property resources, livelihood diversification, or mutual support through networks, and risk avoidance through structural engineering measures or migration (ibid.).

Given the preceding background, a group of representatives from Small Island Developing States in the Pacific (Pacific SIDS) gathered in Suva in August 2013 to discuss high future risks of climate change, with a specific focus on the potential risks of climate engineering (Beyerl and Maas, 2014). The workshop reviewed the current state of scientific research on climate engineering to inform policy discussions surrounding climate engineering research and deployment. A wide range of opinions were raised, including, but not limited to the list of the "Reasons for Concerns" (RfC) about Climate Engineering' (RfC-CE), summarised in Box 24.1.

BOX 24.1 CLIMATE ENGINEERING: PACIFIC SIDS RFC-CE

- **Incomplete knowledge:** Due to still incomplete knowledge about changes in climatic processes' and ecosystems' likely response to these changes, the impacts of climate engineering on global, regional, national, and microclimates bear major uncertainties. Moreover, the subsequent cessation of solar radiation management (SRM) would be particularly difficult. As it does not remove greenhouse gases, it can only suppress the warming for an unknown period of time. Once SRM projects are terminated, temperatures will likely increase at alarming rates, without considerable reductions in greenhouse gas emissions. The warming is likely to be much quicker than present and predicted rates of warming, signifying greater challenges for ecosystems and society.

- **The precautionary principle:** There are enormous ecological, social, environmental, political, ethical, legal, scientific, technological, and economic issues associated with the use and application of climate engineering that are yet to be fully identified and understood. These include questions of procedural and distributive justice, governance, undetermined and unintended consequences of climate engineering on people, systems, and sectors, and national and international legislation to address issues arising from new technologies.
- **False sense of security:** The mere possibility of climate engineering could hinder ongoing international efforts to mitigate and adapt to anthropogenic climate change under the United Nations Framework Convention on Climate Change (UNFCCC) and other international agreements and conventions and related legal instruments, as it may create a false sense of security.
- **Slippery slope effect:** There arises the issue of the "slippery slope effect" from climate engineering research to deployment once climate engineering technologies become available. A slippery slope argument states that a relatively small first step leads to a chain of related events, culminating in some significant effect, much like an object given a small push over the edge of a slope sliding all the way to the bottom.
- **Inclusiveness:** The need for inclusiveness through international and multilateral cooperation as a means of ensuring views and participation of all countries, large or small, developed or developing, on issues such as climate engineering that are likely to affect all human society and natural systems, will be critical for developing locally relevant policies and measures and reducing risks, and,
- **Uncertainties:** Prevailing uncertainty in the sensitivity of the climate system to anthropogenic forcing, inertia in both the coupled climate-carbon cycle and social systems, and the potential for irreversibility and abrupt, non-linear changes in the Earth system with possible significant impacts on human and natural systems, call for research into possible climate engineering options to complement climate change mitigation efforts.

The discussions at the Suva workshop strongly suggest Pacific SIDS are not yet in a position to properly and fully engage in the current climate engineering conversation. While the lack of financial and human resources are the main contributing factors, it is clear Pacific SIDS are again being subjected to an international conversation on a global issue that is likely to have major effects on them and yet have limited knowledge and influence on the outcome.

The current approach to international climate engineering policymaking mirrors that of the climate change policymaking processes in the late 1980s/ early 1990s where Pacific SIDS were subjected to UN-sponsored climate change conferences in the lead-up to the Earth Summit (UNCED) in 1992 without a full understanding of climate change and the stakes to their survival. The experiences have left many Pacific SIDS highly sceptical about externally driven initiatives. Pacific SIDS have to engage in the climate engineering conversation in order to determine regulatory policies and measures in their best interests.

One way of ensuring small islands' perspectives on climate engineering are taken into account is for the international community to establish, without delay, an intergovernmental body, under the auspices of the UN system, tasked with the development of an international governance mechanism for climate engineering research and deployment. Although a number of options have already been explored, there needs to be an internationally sanctioned body to progress development of the governance mechanism.

For Pacific SIDS, one of the key lessons learnt from the climate change negotiations over the past 20 years is the power of working in genuine partnership with other "like-minded" nations under a coalition. Coalition is an effective mechanism for improving a group of countries' bargaining leverage in multilateral negotiations. Pacific SIDS are small and stand to gain little from large UN-sponsored negotiations unless they work together as a coalition.

In 1990, concerned about the effects of climate change and sea-level rise on low-lying islands, a group of like-minded SIDS formed a coalition, the Alliance of Small Island States (AOSIS), to attempt to have a greater impact on UN negotiations in the UNFCCC. Though they differ vastly in their culture, heritage, languages, and economic bases, AOSIS members share a common vulnerability to climate change – the issue that brought them together. AOSIS negotiating positions were guided by a number of carefully formulated principles that pertain specifically to climate policy, including that the international community response to climate change must be: 1) guided by science; 2) that legally binding global GHG emissions reduction targets and timetables commitments be established without delay; 3) recognition of the common but differentiated responsibilities of nations based on their historic emissions; 4) the right to develop; 5) application of the precautionary and polluter pays principles, and 6) fast-track the development and transfer of climate friendly energy efficiency and renewable energy technologies (Lefale, 2001). The most important of these principles from the AOSIS perspective is the second; legally binding global GHG emission reduction targets and timetables commitments, coincidently the same principle underpinning the current rush to fast-track climate engineering research and deployment. The AOSIS principles for climate change policymaking should be adapted to the present dialogue on climate engineering (e.g. CE be (1) guided by science; (2) legally binding policies and measures regulating CE research and deployment; (3) recognition of the common but differentiated responsibilities in managing CE risks; (4) the right to develop CE technologies; (5) CE research and deployment be guided by the precautionary principle; and (6) fast-track the development of an international governance mechanism to regulate the research and deployment of climate engineering techniques, taking into account principles (1) to (5) but more importantly to participate in future international negotiations on the regulation and implementation of climate engineering programmes and activities.

Guidance from the AOSIS and UN development experiences (e.g. Agenda 21, Millennium Development Goals, Barbados Program of Action) should be applied to risk analysis and evaluation of climate engineering in small islands. Potential consequences should be evaluated through an environmental and social impact assessment process. Activities that are considered highly uncertain or with indeterminate outcomes should be carefully monitored. By setting a regulated monitoring system in place at all levels – local, national, regional, subregional, and international – at the early stages of climate engineering research and deployment, potential catastrophes can be minimised.

The challenge for small islands is to ensure their perspectives, needs, and concerns are fully heard and incorporated in the ongoing international climate engineering policymaking dialogue and conversation. Questions about the socio-economic and environmental effects and consequences of climate engineering must be explored before endorsement and implementation of policies and measures that may have irreversible, unintended side effects.

As we have witnessed from the impacts of climate change, global development and progress have predominantly negative consequences for small islands. Mechanisms to encourage their participation and stave off future negative impacts must be financially supported by the international community. We have a history of sidelining small island issues in the global arena, so the question that must be asked about climate engineering policymaking is, "Will their voices be heard this time around?"

First published online as an opinion article in 2014.

25

STRANGE BEDFELLOWS

Climate engineering politics in the United States

Simon Nicholson and Michael Thompson

Climate engineering subverts traditional left–right politics

In the United States, positions on climate change have become strong markers of political identity. There are, in fact, very few stronger indicators. Matthew Nisbet, a political communications scholar, places the current public conversation in the United States about climate change in the same rarefied category as debates about gun control and taxes, as one among a handful of issues that most clearly "show two Americas divided along ideological lines" (Nisbet, 2009). Indeed, a straightforward way to predict whether a particular person in the United States supports action in response to climate change, or, for that matter, believes that climate change is a real thing, is to ask about their political affiliation (for instance, see McCright et al., 2014). Those who identify as Democrats or political liberals are supposed to, as a matter of political dogma, believe in climate change and the need for a response; Republicans or political conservatives are supposed to hold the opposite view.

When it comes to climate engineering, though (and we use the term "climate engineering" in its broadest sense here, to refer to the full potential set of greenhouse gas removal and albedo modification technologies, since this wide usage best reflects the muddied state of the conversation in the United States), the picture is a good deal more complicated. The left–right divide matters, as we will discuss below, but is also subverted and transgressed in interesting and important ways. The growing chatter about development and potential deployment of climate engineering responses is producing curious, and sometimes counter-intuitive, reactions from across the political spectrum.

A few cherry-picked examples give something of the flavour of the entanglements being produced. For instance, a broad coalition of voices across the left–right spectrum stands opposed to climate engineering, at least in public rhetoric.

Al Gore, a liberal champion of climate action, is on record as suggesting that climate engineering, and specifically stratospheric aerosol injection, is "insane, utterly mad and delusional in the extreme".[1] Similarly, author and liberal climate activist Bill McKibben routinely describes any talk of climate engineering as the refuge of those unwilling to shake an addiction to fossil fuels (McKibben, 2013). This broad opposition to consideration of climate engineering is a stance shared with the preponderance of self-identified American conservatives surveyed in a study undertaken by Mercer et al. (2011), the findings of which suggest that conservatives, as a group, skew towards being "detractors" when asked about their views on solar radiation management.

At the same time, a general support for consideration of climate engineering is apparent from interesting quarters. Democratic congressman Bart Gordon, when he was chairman of the House Science and Technology Committee, called the only congressional hearings to date on climate engineering (Geoengineering: Parts I, II, and III, 2010) and is on record as being supportive of the development of a research agenda on many climate engineering technologies.[2] This puts him in uneasy company with Republican political establishment figures like former presidential candidate Newt Gingrich, who made what he called a conservative case for consideration of stratospheric aerosol injection, at the height of debate in the United States about a cap and trade bill, in a blog post with the evocative title, "Stop the Green Pig: Defeat the Boxer-Warner-Lieberman Green Pork Bill Capping American Jobs and Trading America's Future" (Gingrich, 2008).

What's going on? What's shaping these odd coalitions and political boundary crossings?

One possible answer has been advanced by Australian author Clive Hamilton. He has suggested that the divide emerging in the United States between those who claim support for a climate engineering response and those who stand most firmly opposed can best be explained by sharply divergent understandings concerning humanity's use of technology. These understandings says Hamilton, do not track with traditional left–right characterisations. Rather, they reflect worldviews that have their roots in deeper understandings of humanity's relationships with the built and the natural worlds. Hamilton (2013a, p. 18) draws a distinction between so-called *Prometheans* (named for the figure in Greek mythology who wrested fire from the gods) and *Soterians* (named for the Greek goddess who represented, in Hamilton's words, "safety, preservation and deliverance from harm"). The Prometheans hold "[a] technocratic rationalist worldview confident of humanity's ability to control nature", which stands in sharp contrast to the Soterian worldview, "a humble outlook suspicious of unnatural technological solutions and the hubris of mastery projects" (Hamilton, 2013a, pp. 123–124).[3] So Democratic and Republican Prometheans may find common ground in support for climate engineering.

Another contrasting possibility is that climate engineering plays to facets of left–right ideology that stretch beyond traditional conceptions of climate change and what it suggests in terms of required response. Mercer et al. (2011), again, suggest, based on their survey work, that political liberals who stand opposed to climate engineering may do so out of a desire for the maintenance of naturalism when it comes to addressing climate change and other expressions of environmental harm (Wapner, 2013), while political conservatives opposed to climate engineering may be motivated by distrust of government and other powerful institutions that would have most control over a climate engineering response.[4] Or it could be that liberals who advocate for climate engineering research and potential deployment may do so reluctantly out of a sense of desperation when it comes to climate change,[5] while conservatives may believe that climate engineering represents a form of response to climate change that aligns with deep beliefs in the power of free enterprise and market-driven innovation (Lane, 2014).

In this way, the climate engineering conversation can be seen to be creating some strange bedfellows, with pockets on the left and right finding themselves arguing for similar positions, though often for quite different reasons. The result, as Jeff Goodell (2010, p. 15) has put it, is that climate engineering "scrambles old political alliances and carves out new ideological fault lines". There is some intuitive resonance to the kinds of claims advanced by the figures referenced above. They suggest, when taken together, that climate engineering is a subject that cuts across the traditional left–right divide in important and profound ways.

But the left–right political divide still matters

What might these emerging lines of alliance and contestation lead to? We move here from review of the existing landscape in the United States to speculation regarding what it might become.

One important thing to note is that the climate engineering conversation that is evolving rapidly in the United States is doing so in a context that, despite all that has been noted, is marked to its core by left–right ideological understandings. This means that while talk of climate engineering may to some extent confound old categories, it will also be shaped by those categories in ways that must be analysed and certainly cannot be ignored.

This leads to one possible future – a future in which climate engineering is used rhetorically within the United States to subvert climate action. There is a real danger, for instance, that the promise of some climate engineering *deux ex machina* will be dangled as a coming "solution" by conservative political and corporate elites intent on limiting other forms of response to climate change. This is something more than the "moral hazard" that is often discussed in climate engineering circles – that is, the idea that any talk of climate engineering may distract from mitigation and adaptation activities. And it is something more than conservatives embracing climate engineering as a "solution" to a problem that mainstream American political conservatism has long actively denied. Instead, we are suggesting here that it is possible to imagine some vague promise of a climate engineering fix being offered *knowingly* and *strategically* by powerful conservative actors with the sole intent of preserving business as usual. This may happen even as these same actors undercut the kinds of large-scale scientific, technologic, and social investments that would be required to take climate engineering ideas from the drawing board into the world, on the grounds that the public investment and control that, say, a sulphate aerosol injection strategy would require are anathema to conservative ideology.

At the same time, it is possible to imagine climate engineering emerging as a great new bogeyman on the political left. If climate engineering comes to be widely seen or characterised as a conservative plot to avoid taking real action on climate change, then important actors on the political left might rush, in a united fashion, to oppose any climate engineering development. At present, just a handful of environmental non-governmental organisations (ENGOs) are tracking the climate engineering conversation in any meaningful way (Forum for Climate Engineering Assessment, 2015). If real-world events forced ENGOS based in the United States to adopt a position on climate engineering – if, say, a research team were to embark on medium- or large-scale testing of stratospheric sulphate injection, or if a new Republican administration were to announce a desire for a climate engineering research agenda – the traditional left–right divide could well work to prompt fierce liberal opposition.

These are important considerations. The United States has emerged as a hub of climate engineering research, with a growing number of physical science research programmes and an increasingly robust engagement by the scientific establishment. There is also emerging political interest, as evidenced by US government support of a pair of studies produced by the National Academy of Sciences (NRC, 2015a, 2015b) and reports on climate engineering research and governance by important establishment bodies like the Congressional Research Service (Bracmort and Lattanzio, 2013). The positions taken on climate engineering by publics and elites in the United States will, given the country's extraordinary clout when it comes to scientific and technological matters, have an indelible impact on the global climate engineering conversation.

At present, the discussion concerning climate engineering is dominated by scientific voices, speaking, for the most part, in rationalist ways about climate engineering as, at best, a bottom-of-the-barrel component of a strategic response to climate change. However, when climate engineering gains more political and public traction and salience in the United States, as it surely must, the terms of the conversation will change. Broader climate politics in the United States are anything but rational and are instead driven all too much by political ideology and culture. So, too, will it be for the politics of climate engineering.

First published online as an opinion article in 2015.

Notes

1 Quoted in Goldenberg, 2014. Gore, who acts as a judge for the Virgin Earth Challenge, is said to support research and development of many proposed carbon dioxide removal technologies, some of which traditionally fall under the climate engineering umbrella. He was specifically referring to stratospheric aerosol injection in his comment cited above.

2 See, for instance, Gordon's comments as part of a panel discussion on climate engineering organised by Resources for the Future and the Forum for Climate Engineering Assessment, available at http://www.rff.org/Events/Pages/Whats-Next-for-Climate-Engineering.aspx. Gordon argues that various climate engineering technologies may well, with United States government support, become "tools in the toolbox".

3 By Hamilton's reckoning, the Promethean (or what Hamilton calls in Revkin, 2014 the "eco-pragmatist") camp is occupied, in the United States, by figures like Michael Schellenberger, Ted Nordhaus, Peter Kareiva, and Stewart Brand.

4 However, see Buck, 2015, for an argument that this simplistic account of conservative motivations breaks down in the face of the fact that deployment of large-scale climate engineering technologies would require a strong overarching regulatory scheme.

5 See, for instance, the profile of Ken Caldeira in Merchant, 2014.

26

GEOENGINEERING AND THE HUMANITARIAN CHALLENGE

What role for the most vulnerable?

Pablo Suarez, Bidisha Banerjee, and Janot Mendler de Suarez

As research practitioners working on climate and disasters in the humanitarian sector, we are at once fascinated and terrified by the prospect of geoengineering. Rapidly becoming technically feasible as a "planetary emergency procedure" (Kintisch, 2010b) – somewhere, somehow, the intentional manipulation of the global climate may become politically feasible as well during our lifetime. Sullivan argues that to the engineer, Murphy's Law ("what can go wrong will go wrong") represents a statistical truism (Sullivan, 1995); if deployed – geoengineering can go wrong, and *will* go wrong in some way.

Altering the Earth's climate, whether inadvertently through anthropogenic greenhouse gas emissions or deliberately through geoengineering, is an experiment in which every person on our planet is potentially a test subject (Suarez et al., 2010). Vulnerable populations will be differentially and disproportionately impacted by deployments. Methods like solar radiation management (SRM), i.e. deflecting sunlight through the dispersal of sulphur particles in the upper atmosphere, can cool the planet (the desired effect), but – like global climate change – can also trigger changes in rainfall patterns that are nearly impossible to predict.[1] Haywood and colleagues suggest that volcanic eruptions, which SRM would mimic, influenced increased desertification of the Sahel between 1970 and 1990 (Haywood et al., 2013). This is the Sahelian drought in which 250,000 people perished, leaving 10 million refugees in its wake. While they conclude, "Further studies of the detailed regional impacts on the Sahel and other vulnerable areas are required to inform policymakers in developing careful consensual global governance before any practical solar radiation management geoengineering scheme is implemented", new geophysical research suggests a causal correlation between these Sahelian droughts and sulphate aerosols from coal-burning in North America and Europe (Hwang et al., 2013).

This raises two key questions from the humanitarian perspective:

i.   What role, voice, or agency will the vulnerable have in geoengineering decisions?
ii. Who will pay for humanitarian operations in a geoengineered future?

Regrettably, past experience and Murphy's Law invite two unacceptable predictions:

The most likely answer to the first question is simply, "None". There is currently no governance framework for geoengineering research or deployment, scarce evidence of any real effort to include the really vulnerable in the rarefied debate around whether or not to act, and it is generally deemed too expensive to seek informed consent from all those at risk should the experiment go awry. As argued by Blackstock and Long (2010), geoengineering stakeholders need to consider whether existing frameworks can facilitate an accessible, transparent process, or whether new fora, treaties, and organisations are required.

For the second question, the status quo answer is "Nobody who causes harm through geoengineering will pay for humanitarian operations", regardless of whether they constitute "normal" disaster management or humanitarian aid directly in response to geoengineering impacts. Geoengineering deployments that shift the burden of impacts constitute humanitarian externalities. The history of the United Nations Framework Convention on Climate Change suggests that institutional forces allow existing greenhouse gas externalities to prevail. This inertia can only be expected to become more entrenched when the climate system shows more signs of man-made instability.

Notwithstanding, we need to prove these dastardly predictions wrong. To ensure that the vulnerable have increased input into decision-making processes, and that anticipating and paying for the negative impacts of geoengineering is "internalised" by deployers, it is crucial to begin identifying principles, pathways, and structures that might address concerns from the humanitarian community and the differentially vulnerable populations they serve.

Proponents ask a valid question: will the results of deploying geoengineering be worse than the alternative – inaction in the face of accelerating global climate change? The answer is both "We don't know" and "It depends". Humanitarian work is precisely about addressing things gone wrong. Victims of disasters are often the victims of individual and collective failures to reduce disaster risks, and the more vulnerable are differentially impacted. A drought that kills thousands in Niger may only lead to reduced profits if in Nebraska. The IPCC's Special Report on Extreme Events (2012) asserts that a new balance needs to be struck between measures to reduce risk, transfer risk (e.g. through insurance) and effectively prepare for and manage disaster impact in a changing climate (IPCC, 2012). This balance will require strengthened emphasis on *anticipation* and *risk reduction*. Given the prospects of geoengineering, it is our duty to anticipate what can go wrong for those who lack the means to cope with surprises and to ensure that the most vulnerable have the capacity to effectively manage anticipated risks.

We posit that differential impacts on the more vulnerable are highly probable in a geoengineered climate, so if decisions are made to deploy SRM for the intended benefit of one portion of the global population while causing others to suffer, this intentional shifting of the burden of coping represents a humanitarian externality.

We propose a guiding framework to examine the humanitarian challenge based on the concept of negative externality, i.e. when party A (e.g. a developed nation[2]) seeking benefit B (i.e. improved climate) implements an activity or transaction T (e.g. solar radiation management) with consequences which cause losses or costs C (e.g. crop failure) to an otherwise uninvolved party Z (e.g. subsistence farmers in a particular developing nation or nations), who did not choose to incur the negative impact.

The prospect of insufficient emissions reduction is irrefutable, largely due to the inability of governance and market structures to address such externalities. As Martinez-Alier (2002) has argued, the notion of an externality as a "market failure" can also be understood as a successfully transferred cost. Some benefit, many others suffer. With global warming already unavoidable, we face the humanitarian imperative of addressing its consequences: even as mitigation and adaptation remain critically important, and interest in geoengineering increases, it is important to remember that we have the collective knowledge and capability to create paradigm-shifting ways to address humanitarian concerns.

We propose framing geoengineering research and policy agendas in ways that explicitly integrate the role of the most vulnerable through *Learning*, *Preparing*, and *Preventing* as a way to internalise humanitarian externalities:

Learning: Identify Z's likely costs C, and research them at least as much as benefits B are researched by Party A. (In other words, we should, for example, identify the subsistence farmers' likely costs due to crop failure as a result of SRM and research these costs as much as benefits (such as climate stabilisation) are researched by the developed nation where SRM research is taking place.) This suggestion is particularly directed to climate scientists.

Preparing: Identify the party Z (e.g. the subsistence farmer) who would be excluded from decision-making about activity T and seek to include her. Importantly, help Z understand decision processes potentially leading to activity T, so as to confer agency and capacity to influence her own future. If the risk of negative consequences is deemed high, support her in reducing likely costs C (e.g. crop failure). This suggestion is particularly directed to policymakers and the humanitarian sector.

Preventing: Seek to establish mechanisms that either prevent the successful transfer of cost C (e.g. the cost of crop failure) from any party A (e.g. the nation deploying SRM) to any party Z (e.g. subsistence farmers), or internalise the externality (through mechanisms such as environmental assurance bonds – described below).

For *Learning*, we can design active participatory research that foregrounds the interests and involvement of the most vulnerable, to broaden the range and relevance to them of the research questions. This includes research into the more nuanced aspects of atmospheric behaviour that play determining roles in the lives and livelihoods of subsistence farmers, shantytown dwellers, and others already on the edge of survival. For example, maize cultivation in southern Africa critically depends on the timing of precipitation within the parameters of the crop's phenological cycle. The same applies to the incidence of unusual conditions conducive to crop pests. Thus food security is very sensitive to changes in atmospheric circulation patterns that drive changes in precipitation such as the onset and duration of seasonal rains, frequency and timing of dry spells, or rainfall intensity during critical periods such as germination and ripening (Tadross et al., 2009). However, most geoengineering impact modelling examines less nuanced aspects, such as mean annual precipitation or temperature changes (see for example Royal Society, 2009). What such research can offer is to uncloak our colossal ignorance about the likely negative impacts of geoengineering – and help to reduce this knowledge gap.

For *Preparing*, geoengineering stakeholders should support new approaches to participatory decision-making and expand the scope of how we create and share knowledge about what is possible – and desirable. Options include experiential games for learning and dialogue that simulate the complexity of climate-related decisions and related consequences, to help engender systems-thinking including information needs, feedbacks, delays, thresholds, and trade-offs (Mendler de Suarez et al., 2012),[3] as well as participatory video – a way to involve a group in filming their own story, from storyboarding to interviewing and camera operation that enables people to distil and share their own experience and insights (Suarez et al., 2011).

Of course, experiential games and video in and of themselves may not be enough to support party Z (e.g. the subsistence farmer) in reducing likely costs C (e.g. crop failure). Nevertheless, participatory decision-making can spark a deeper discussion amongst all levels of society about how best to support party Z. If people are invested at all levels, we are more likely to find useful answers and perhaps formulate a formal framework. As of now, absolutely nothing is being done to let party Z know what geoengineering proponents are contemplating. These creative platforms may enable the most vulnerable to form an opinion and perhaps influence formal governance processes. It is the government's responsibility to elicit party Z's opinion, which brings up the usual challenges associated with bringing in the voices of the marginalised. But, at least, thanks to these participatory mechanisms, party Z will have an informed opinion – and a motivation to make that opinion count.

For *Preventing*, it is necessary to open and nurture an inclusive process for examining key governance decisions and research areas so that conflicting definitions or designations of vulnerability and geoengineering-related impacts, loss, and damages may be reconciled. Additionally, we should support research and policy efforts to innovate through corrective mechanisms and financial instruments designed to internalise the negative externalities incurred by the most vulnerable. Possibilities range from international governance frameworks to further research into market-based tools. The development of environmental assurance bonds, which could require geoengineers or their funders to post a guaranteed price equivalent to the worst-case threats posited by a particular deployment scheme, looks promising (Banerjee, 2011).

At present, from the humanitarian perspective it is not possible to reliably compare the risks of geoengineering to the risks of unfettered climate change. It is important to note that we are not experts in the many disciplines needed to inform an agenda and policy on their relative impacts and externalities. The growing geoengineering community is actively exploring options. We submit that there is a moral imperative to facilitate involving the most vulnerable in decision-making about geoengineering. The concept of internalising potential negative externalities offers an operational framework for applying the principles and pathways suggested in this article, to help inform a more inclusive and nuanced conversation about what can go wrong – and what must go right.

First published online as an opinion article in 2013.

Acknowledgements

We are grateful to the many people and organisations that invite inclusion of the humanitarian perspective in ongoing processes examining geoengineering concerns, including the Asilomar Conference on Climate Intervention Technologies, the many events of the Solar Radiation Management Governance Initiative (SRMGI), the Geoengineering Scenarios Workshop at Yale, and the "Geoengineering Our Climate" conference in Ottawa. The views expressed in this opinion piece are solely those of the authors and do not represent the positions of their organisation.

Notes

1 It is perhaps equally challenging to predict local impacts of climate change. Moreover, it is unknown how intentional climate modification would interact with short-lived climate forcers like black carbon from fossil fuel burning.
2 There is also the possibility that a developed nation could seek to deploy geoengineering (probably SRM) as a humanitarian measure on behalf of a vulnerable nation. (See Suarez et al., 2010.)
3 Participatory games have successfully been used to engage illiterate Ethiopian farmers in the design of complex index insurance instruments and other climate risk financing approaches – they surely can help engage humanitarian workers and vulnerable people in the geoengineering debate.

27

OPPOSITION TO GEOENGINEERING

There's no place like H.O.M.E.

Action group on erosion, technology, and concentration (ETC Group)

In April 2010, the World Peoples' Conference on Climate Change and the Rights of Mother Earth, held in Cochabamba, Bolivia, brought together more than 25,000 *campesinos*, teachers, students, engineers, activists, diplomats, elders, and ordinary folk to discuss how best to minimise the impacts of global warming and to respond to the failure of negotiations at the UN Framework Convention on Climate Change to bring about reductions in global greenhouse gas (GHG) emissions. Seventeen working groups contributed to a Peoples Agreement, which explicitly rejected geoengineering as a "false solution" to climate change.[1] From Cochabamba, the "Hands Off Mother Earth" (HOME) campaign to oppose geoengineering experiments was launched.[2]

Since Cochabamba, a small but influential group of researchers has increased calls for governments to support geoengineering experiments as part of developing a "Plan B" (Connor, 2009) or "insurance policy" (Alleyne, 2009) in the event of a "climate emergency" (Caldeira and Keith, 2010) – despite the adoption of a decision to restrict geoengineering activities by the UN's Convention on Biological Diversity (CBD) in October 2010.

The CBD is an international legal instrument whose aim is the conservation and sustainable use of biological diversity. Some 193 countries are parties to the convention (only the Holy See, Andorra, South Sudan and the United States are not). At its Conference of the Parties held in Nagoya, Japan in 2010, the CBD extended a decision that restricts one specific geoengineering technique agreed upon in 2008 (on ocean fertilisation) to apply to all geoengineering activities, while allowing small-scale scientific research studies that meet certain criteria.[3] For many states, the CBD meeting provided the first opportunity to discuss the kinds of geoengineering technologies under development and to consider their risks.

The push for Decision X/33 at the CBD, which we argue is a *de facto moratorium* on geoengineering, came largely from governments of the global South – including the African Group and ALBA countries as well as coastal countries such as Philippines and Tuvalu, the majority of whose peoples rely on oceans for their livelihoods (IISD, 2010). Sunlight-reflecting stratospheric sulphate injections are expected to alter precipitation patterns, particularly in the tropics, and ocean fertilisation techniques deliberately alter ocean chemistry in an attempt to increase absorption of carbon dioxide.

Some hailed the CBD's decision as a prudent and necessary measure until international regulations can be developed and impacts can be properly assessed, while others questioned the decision's relevancy or enforceability (Kintisch, 2010c; ETC Group, 2010c). *The Economist* (2010) saw the UN's attention to geoengineering as the first chapter in its "coming of age" story. HOME campaigners interpreted it as a stopgap until a global ban on any unilateral attempt to engineer the climate can be negotiated. In any case, the CBD decision signalled the first baby steps toward intergovernmental regulations – something that has been opposed by advocates of geoengineering research such as those attending the Asilomar International Conference on Climate Intervention, who expressed their preference for a voluntary system of self-governance. The explicit goal of the conference, organised by the Climate Response Fund was to suggest precautions to assure safe conduct of experimentation and to "propose voluntary standards for climate intervention research" (Asilomar, 2010).

From some perspectives, geoengineering as "Plan B" or "insurance policy" in the event of a "climate emergency" may seem prudent, practical, and even precautionary. But geoengineering's prudence will not be universally obvious. If you are the G-8 member that launched the Industrial Revolution causing climate change and your GHG emissions keep going up instead of down – it may be easier to appreciate the attraction of a "techno fix". As it is likely that only the world's richest countries will be able to develop the hardware and software necessary to reset the global thermostat, it will be the governments that are responsible for almost all historic GHG emissions and have either denied or ignored climate change for decades, which will also have de facto control over the deployment of geoengineering experiments. Those same governments have failed to provide even minimal funds for climate change mitigation or adaptation. It defies reason to suggest, as some have, that geoengineering will not divert funding and intellectual resources from mitigation and adaptation; it already has – the UK's Royal Society, the American Academies, the IPCC, to name a few, have all spent money and time bringing experts together to consider geoengineering's prospects.

Further, to have an impact on the Earth's climate, geoengineering projects will have to be on a massive scale. Projects that alter the stratosphere or the oceans will not only have unknown implications but also unequal impacts across the globe

(Royal Society, 2009, p. 62). As much as the unintentional "geoengineering" of the Industrial Revolution disproportionately harms tropical and subtropical parts of the planet, intentional geoengineering experiments could well do the same. Put bluntly, many South governments lack a blind faith in technology to solve the problem of climate change, and, in our view, a lack of trust in the governments, industries, or scientists of the North to protect all the world's people is justified. In the absence of demonstrable goodwill and humility from the governments likely to conduct geoengineering, it would only be sensible for the peoples and governments of the global South to be suspicious.

A rejection of geoengineering is not a denial that science has an important role to play in dealing with climate change. It is urgent and important that the scientific community work with national and even local governments monitor and address the climate threats ahead. This collaborative effort will require a lot of money and focused energy. The practical responses to climate change must change with the latitudes, altitudes, and ecosystems.

"Hands Off Mother Earth" campaigners assert that not enough is known about the Earth's systems to risk geoengineering experiments in the real world. No one knows if these experiments are going to be inexpensive, as is often assumed – especially if they don't work, forestall more constructive alternatives, or cause adverse effects. We don't know how to recall a technology once it has been released. Beyond those uncertainties and inadequacies, we must acknowledge the geopolitical realities of climate change. Without that acknowledgement, geoengineering can only be *geopiracy* and it is a threat to the entire natural world, including each one of us calling Earth *HOME*.

First published online as an opinion article in 2013.

Notes

1 The Peoples Agreement is available online: http://pwccc.wordpress.com/support/.
2 The HOME campaign specifically opposes open-air geoengineering experiments – computer modelling and contained laboratory tests are not targets of the HOME campaign: www.handsoffmotherearth.org.
3 Decision X/33, para 8 (w): "Ensure, in line and consistent with decision IX/16 C, on ocean fertilization and biodiversity and climate change, in the absence of science based, global, transparent and effective control and regulatory mechanisms for geoengineering, and in accordance with the precautionary approach and Article 14 of the Convention, that no climate-related geo-engineering activities that may affect biodiversity take place, until there is an adequate scientific basis on which to justify such activities and appropriate consideration of the associated risks for the environment and biodiversity and associated social, economic and cultural impacts, with the exception of small scale scientific research studies that would be conducted in a controlled setting in accordance with Article 3 of the Convention, and only if they are justified by the need to gather specific scientific data and are subject to a thorough prior assessment of the potential impacts on the environment".

Footnote to decision X/33 para 8 (w): Without prejudice to future deliberations on the definition of geo-engineering activities, understanding that any technologies that deliberately reduce solar insolation or increase carbon sequestration from the atmosphere on a large scale that may affect biodiversity (excluding carbon capture and storage from fossil fuels when it captures carbon dioxide before it is released into the atmosphere) should be considered as forms of geo-engineering which are relevant to the Convention on Biological Diversity until a more precise definition can be developed. It is noted that solar insolation is defined as a measure of solar radiation energy received on a given surface area in a given hour and that carbon sequestration is defined as the process of increasing the carbon content of a reservoir/pool other than the atmosphere.

28

IS SOLAR GEOENGINEERING A NATIONAL SECURITY RISK?

Chad Briggs

A growing number of discussions on geoengineering risks and politics pose the question of how solar radiation management (SRM – or other means of abrupt climate forcing) might be of interest to security and military planners. Scenarios have also introduced discussions of deliberate actions to disrupt the climate for military or terrorist goals, and/or military responses to geoengineering policies.[1] Despite the history of military attempts to control weather and environmental conditions (Fleming, 2018), it is important not to simplify geoengineering security concerns or to overstate potential military responses.

Climate and environmental changes affect security at both the strategic and operational level but do so by potentially destabilising underlying, vulnerable systems (whether ecological, economic, geophysical, or social). Like cyber security, there are no clear thresholds for what constitutes a "threat", or what a proportional response might involve. Many militaries have a keen interest in mapping out potential cascading impacts from climate change as they affect security missions, but in general these risks are seen as indirect, and that the role of planning is to adapt to potentially shifting conditions (CNA Corp., 2007). This perspective briefly describes two security-related points within solar radiation management: that SRM does pose security risks that are of interest to military planners and may result in security impacts, but that within the US security community the issue is so legally and politically sensitive that the US Department of Defense has no interest in pursuing concrete actions in this field.

SRM can produce security-related impacts for the same reasons that climate change is considered a "threat multiplier" in the 2010 Quadrennial Defence Review (QDR), in potentially destabilising social, economic, and environmental systems upon which state security relies. National security consists not only of planning for direct military actions between states but is a much broader concept that takes into account when background conditions (in military war-game vocabulary, "environmental" systems) change with indirect but unexpected results for international or state stability. For example, although Somalia poses little direct threat to the United States, instability in the region can be considered an accelerant for increases in terrorism, piracy, and human-related disasters (e.g. famine), risks which the US security community must address at one level or another.

SRM contains two differences: first, impacts from SRM are the result of intentional human actions to change environmental conditions; second, the effects of very large deployments are intended to manifest quickly. SRM also shares one parallel: even the deployment of methods that aim primarily to cool a particular region (e.g. marine cloud brightening) will still have knock-on effects on wider environmental and human systems. Indeed, the deployment of other SRM techniques, such as sulphate particle injections, will have necessarily global impacts. These characteristics are significant for gauging the perspective of security planners.

Abrupt environmental changes are far more worrying from a security perspective than gradual change, as abrupt changes are less predictable, offer less chance for systems to adapt and/or recover from changes, and are more likely to result in disaster situations. Disasters in this perspective refer to combinations of events that overwhelm vulnerable points in related systems, and where resilience is not high enough to prevent negative, cascading impacts that can spread far from the original disaster. The 2011 Tohoku earthquake, for example, revealed that general urban resilience of Japan was high enough to cope with the magnitude 9.0 quake, but a critical vulnerability in the Fukushima Daiichi nuclear power plant (the backup diesel generators) resulted in a disaster with cascading impacts not only on Japan and its economy but energy flows and systems across the globe. Likewise with local flooding events, a city can cope with a certain amount of rainfall averaged over months, but receiving the same amount in a short period of time generally results in disasters (and where the military is often called to provide relief).[2]

The security concern with SRM is therefore that related heating or cooling will result in non-linear stresses on environmental systems, where vulnerable parts of the system will be damaged or collapse, resulting in cascading impacts which may in turn destabilise vulnerable social, political, and economic systems which rely upon a steady-state environment. As these systems are often globalised, failure in one area may create cascading impacts. An attempt to cool, for example, glacial highlands in Peru may unintentionally result in non-linear shifts to precipitation patterns in the Amazon basin of Brazil, with second- and third-order effects on rainforest health, food security, and the global carbon cycle.

Despite the difficulties mentioned above, military planning does offer some guidance as to how geoengineering might be considered a security risk, and how to assess such complex risks. If direct causality is difficult to determine, groups can still conduct comprehensive risk assessments from environmental changes and cascading impacts on complex systems. Rather than take a traditional security view of militaries and violent conflict (in military terms, "kinetic operations"), the focus is on stability and well-being of communities and the ecosystems and socio-economic and infrastructural systems upon which they rely. And in contrast to scenario methods that either combine two driving forces into four alternative futures (Shell or GBN), or which combine a set of drivers and projects into the future (RAND or Hudson Institute), new energy-environment scenarios in the US military do not attempt to describe a complex world of social, economic, political, and technological changes (Kahn and Wiener, 1967; Wack, 1985).

Instead of traditional scenario methods, security planners can take a given set of starting conditions (often geophysical variables, set five to ten years in the future), and recognise that the affected systems themselves already contain significant complexity. By tracing multiple pathways of impact/response, expert groups can map out alternative futures for a given region. These futures can then help identify where additional monitoring is needed, where critical intervention points might exist, and where secondary or tertiary impacts create security concerns. For example, the loss of meltwater from Tibetan glaciers is not itself a security risk. But the significant impacts and resulting actions (downstream flows to South-East Asia, loss of food security and energy resources, etc.) can spark humanitarian or political crises requiring outside intervention (Briggs, 2010).

The US military considers environmental conditions to be important background considerations for long-term strategic and short-term operational planning; yet despite its desire for such systems to remain stable over time, this does not translate into any desire to pursue SRM-related projects.[3] The first reason for this is the highly questionable legality of military-related actions designed to affect climate, as reflected in both internal and domestic regulations, the 1977 Convention on the Prohibition of Military or Any Other Hostile Use of Environmental Modification Techniques (ENMOD), and the 2010 Convention on Biological Diversity.[4] ENMOD specifically forbids weather or climate modification for hostile purposes. Though the definition of "hostile" in the context of SRM is complex, direct aggressive military use would certainly qualify (Reynolds, 2015). Politically, these restrictions follow decades of negative experiences in attempting to alter environmental conditions during/around conflict, most notably USAF's attempt to modify weather during the Vietnam War (Fleming, 2010). The US and Soviet Union adopted ENMOD shortly after the Vietnam War to prevent future military applications of weather and climate modification.

A general view in the US security community holds that geoengineering (and in particular, SRM) projects are not tightly controllable, and therefore are analogous in some ways to biological or chemical agents. Militaries tend to dislike platforms that are not tightly controllable and predictable, one reason why biological weapons were used so rarely in history (Steinbruner, 1997). Although a disease like weaponised smallpox could easily defeat one's adversaries, officers understand that viruses do not discriminate and that casualties may spread very far from the intended targets, including their own troops and civilians.

Security planners in the US traditionally also rely upon a concept called "proportional response" in planning, where intentional attacks upon one side are met with a similarly destructive (or restrained) response from the other. With general climate change there is no intent involved to change the environment, and therefore climate change exists as a background risk, and no direct agency is involved. Yet if the change is intentional, what then is the appropriate response? And if SRM technologies are used by one state against another for with hostile intent, what are the thresholds for response, and by what means? The intentionality of action question is further complicated in cases where corporations or private individuals (and not states) are the actors (US Air Force, 2000).

As long as militaries remain uninvolved in geoengineering (and there is nothing to say that some countries may not try), SRM is not a *prime facie* security risk, and the US military will likely steer clear of the issue for at least the near future. However, any actions taken to alter the climate may have significant impacts that could quickly translate – even unintentionally – to legitimate security concerns. Research communities can help to identify potential risks well in advance, by establishing what changes might be possible with given SRM technologies, and what cascading impacts this would have on a given region.[5] This effort can in turn help discussions of necessary governance and research, by identifying both potential impacts and critical uncertainties. Given the near-irreversible nature of environmental changes, a good measure of caution and foresight should easily be justified.

First published online as an opinion article in 2013.

Notes

1 For example, during the SECURENV scenario workshop in Stockholm, Sweden, November 3–4, 2010.
2 The concept of fragility was taken from engineering concepts, and refers to a system's risk of "phase shift", or reverting to a different and perhaps lower order of stability. (See Wisner et al., 2005.)
3 In fact, during this author's time directing the USAF Minerva programme on energy and environmental security, geoengineering was the one area deemed too sensitive to include in the team's research portfolio.
4 See: http://www.icrc.org/ihl.nsf/INTRO/460
5 For methodological reasons, global assessments lack necessary detail and resolution, requiring impacts to be assessed regionally.

PART VI

Geoengineering governance

From research to the real world

Introduction

This final section of the book is dedicated to looking forward. We suspect that most of geoengineering's history, and certainly the most interesting parts, have yet to be written. As such, the contributions throughout this section reflect upon ways to explore and manage geoengineering as an emerging issue: methods by which the "future" be engaged with, and mechanisms and templates for upstream governance. Indeed, if there are multiple degrees of freedom in modelling the climatic repercussions of deploying geoengineering approaches, then gauging and governing its political dimensions is an exercise in even greater uncertainty.

The first four chapters focus on the governance of geoengineering research, reviewing principles and mechanisms for guiding intent and action on the part of scientists and technologists, for generating funding and collaborative networks, and for engaging incoming stakeholder groups. David Morrow and colleagues begin in Chapter 29 with a proposal for a set of ethical principles to guide geoengineering research based on similar principles that have applied to research involving human and animal subjects for many decades. These principles have a strong resonance with many of the Oxford Principles for geoengineering research, for which Tim Kruger provides a history and commentary in Chapter 30. In Chapter 31, Arunabha Ghosh rounds out this set of papers with a proposal for the operational design aspects of an international geoengineering research programme. Within the chapter, Ghosh first derives these design aspects from the governance challenges facing geoengineering research and then examines a host of international research programmes on different topics for mechanisms that might successfully address them. In Chapter 32, Alex Hanafi and Steven Hamburg describe lessons from international workshops and networks exploring the governance of geoengineering convened by an influential set of non-governmental actors as part of the Solar Radiation Management Governance Initiative (SRMGI). Based on these lessons, they argue articulately for "inclusive and adaptive governance of SRM research" and suggest continuing integrated, transnational conversations to avoid "isolated dialogues [that] risk wasting limited resources while failing to build the cooperative bridges needed to manage [future] international conflicts".

These contributions reflect wider movements in the field. Much of the early literature on governance (immediately after 2006) focused on the management of risks stemming from deployment and identified international legal frameworks as appropriate sites of governance (see Part IV, Reynolds and other legal reviews). This was supplemented by a widespread view that the main governance challenge was constraining over-eager scientists and unilateral actors in the private sector, or at the state level by individual states or coalitions of the willing. In intervening years, the considerable caution exercised by the research community, coupled with halting movements on field experimentation or pilot projects in both CDR and SRM, have led to shifting perceptions that there are currently few incentives for developing forms of geoengineering and that governance proposals should reflect this reality. As a result, we have seen an enlarged focus on less formalised, non-legal mechanisms at the level of research programmes and communities, and on pre-deployment phases of action – research enablement and funding, field tests, stakeholder engagement, and future-oriented risk exploration. Some recent publications might be indicative. Nicholson et al. (2017) provide an admirable summary of these evolutions of thought and a call for "polycentric" forms of governance in SRM. A report by the Solar Radiation Management Governance Initiative (2016) provides what is still the most comprehensive typology of research governance, while Oschlies et al. (2016), as part of the German National Priority Program for assessing geoengineering research, generate a set of decision-oriented metrics and criteria that emphasises the incorporation of multiple disciplines and stakeholder groups.

Johannes Gabriel and Sean Low pick up these themes in Chapter 33 for the angle of using foresight to enabling collective, repetitive learning about the evolving risks and opportunities of geoengineering technologies and governance. Building from two international examples of foresight exercises on geoengineering technologies, they conclude that foresight may be "an approach that could help society explore [geoengineering's] many different pathways and outcomes". Such explorative scenario exercises have become increasingly common recently as tools for exploring a complex range of future actions and contingencies. Proponents note their potential to structure communication, in a participatory and experimental fashion, between myriad stakeholders and bodies of disciplinary knowledge that currently exercise forceful claims upon how an engineered climate might impact human and natural systems (Low, 2017a,b; Bellamy and Healey, 2018; Vervoost and Gupta, 2018). Such forms of foresight form an integral part of "anticipatory" frameworks for the governance of emerging technologies that are being applied, not only in geoengineering, but in novel and controversial fields of technology development with applications in medicine, warfare, mobility, and industry.

This is the subject of the contribution from Rider Foley and colleagues, providing the last word in this book. Building on the themes of inclusiveness and foresight raised in the preceding chapters, in Chapter 34 Foley et al. present the framework of anticipatory governance as a potential model for governing the emergence of geoengineering technologies. Though applying their framework to the analysis of five diverse reports on geoengineering technologies and governance, they identify "gaps [within the] various approaches to foresight, engagement, and integration" of these reports, that increased attention to inclusive and adaptiveness within the geoengineering discourse could address going forward. Yet, this is but one particular application that treats research itself as a form of governance. One might note that an upstream governance framework applied to a prominent attempted field experiment in SRM (the SPICE test bed, see Doughty, Chapter 15) was fashioned on precepts of anticipatory governance – or, in the terminology of a much-related field, responsible research and innovation (Stilgoe et al., 2013a). These frameworks place a spotlight upon the foundational role of scientists and technologists in setting the terms of debate, and provide guidelines for how to proactively explore and manage concerns in real time through modes of paced, reflective, and democratised scientific deliberation (Stilgoe, 2015).

into the atmosphere to observe their behaviour, and research in carbon dioxide removal technologies, do not pose the same risk as large-scale SRM trials. Thus, these principles do not necessarily apply to them.

These principles derive from well-established ethical principles for research with human and animal subjects. Before beginning research involving human subjects, researchers need approval from their institution's Institutional Review Board (IRB).[1] IRBs grant expedited clearance to studies that pose minimal risk, but they require closer scrutiny of studies that pose more than minimal risk or involve "vulnerable populations" (i.e., persons who might not be able to defend their own interests). Ethicists identify three conditions that such research must meet. Researchers must convince their IRB that a study meets these three conditions before it may begin. Since large-scale SRM trials would put millions of people at risk for the sake of producing knowledge, they should meet similar conditions.

The first condition of ethical research, on which we base our Principle of Respect, is that all subjects must participate in the research willingly and with full knowledge of the nature and risks of the study (National Commission, 1979). No one may be coerced or manipulated into participating. This enables individuals to exercise autonomy in deciding what happens to them and what risks to take. Traditionally, this principle is implemented by requiring each participant's informed consent. Since SRM experiments are collective decisions, the relevant standard is not universal consent but a legitimate decision made by a politically legitimate institution representing all affected persons (Morrow et al., 2013). To say that an institution is politically legitimate is to say that it has the moral authority to make collective decisions on behalf of its constituents. A particular decision is a legitimate decision only if it is made by a legitimate body and meets certain substantive conditions, chief among which is respecting the individuals who are affected by the decision (ibid.). For example, a decision to allow unsafe dumping of radioactive wastes near a town inhabited by an unpopular minority group would be an illegitimate decision, even if it were made by a legitimately elected legislature. Since a large-scale SRM trial would affect persons in many countries, only an international body or agreement would have the political legitimacy to authorise the experiment. Such authorisation would constitute a legitimate decision only if it adequately respects and protects those who would be endangered by it. Identifying or designing an effective governance institution that has moral authority over large-scale, risk-laden GE experiments is a theoretically and practically daunting task, beyond the scope of the present paper.

This dovetails with the second condition of ethical research, which is that risks and benefits be appropriately balanced (National Commission, 1979). This is especially difficult to implement in large-scale SRM trials since risks and benefits will be distributed unevenly across regions, economic sectors, etc. Some people might face risks for which they receive no offsetting benefit, especially if the benefits from research come decades later. One partial response is to establish a fund to compensate those who suffer climate-related harms during an experiment.[2] Another is to choose experiments whose anticipated risks fall on those best able to bear them, although much work remains to be done in figuring out how to implement that guideline (Morrow et al., 2009).

The third condition of ethical research is that all participants must be treated justly (National Commission, 1979). In particular, the burdens of risk-laden research may not be shifted unfairly to those least able to refuse participation, and the benefits of research must be shared with those who undertook the risks. For GE research, this condition requires distributing expected benefits and risks of GE experiments justly. For instance, a country sponsoring an SRM trial should not deliberately structure the experiment to shift the risk to other countries. The exact principles by which expected benefits and risks should be distributed – to the extent that this can be controlled at all – still need to be worked out (Morrow et al., 2009). Conventional research ethics provides little guidance here; decision-makers should look to political philosophy. Our Principle of Beneficence and Justice derives from these second and third conditions for ethical research.

Our Principle of Minimisation derives from the ethics of animal research, where the "replacement, reduction, and refinement" of animal use guides experimental design. In the GE context, this slogan implies that experiments should be as small (in geographic scale), as short (in duration), and as non-disruptive (in terms of climatic and environmental impact) as is necessary to test specific scientific hypotheses. To be scientifically useful, SRM trials would have to be global, multi-year interventions that change the climate in noticeable ways, but they should not be any longer or more intense than necessary (Morrow et al., 2009).

The three principles articulated here complement the Oxford Principles for GE governance (Rayner et al., 2013). The Oxford Principles address both GE deployment and research. Where they address large-scale, risk-laden GE research, the Oxford Principles cohere well with our three principles. Our Principle of Respect requires politically legitimate institutions, which, we argue elsewhere (Morrow et al., 2013), would need to be transparent and accountable to the public; this requirement coincides with the Oxford Principles' requirements for public participation in decision-making, the disclosure of research results, and independent impact assessments. Our Principle of Beneficence and Justice and our Principle of Minimisation add further ethical constraints on large-scale SRM trials – constraints that do not conflict with the Oxford Principles.

A great deal of conceptual and political work remains to be done to operationalise these principles and create the institutions needed to apply them. The need for regulating GE research is coming faster than the need to regulate GE deployment. Fortunately, the existing framework for regulating other research provides a model for regulating GE research.

First published online as an opinion article in 2013.

Notes

1 A basic introduction to research ethics is freely available through the US National Institutes of Health at http://phrp.nihtraining.com.
2 Such compensation schemes face serious challenges, however. See Svoboda and Irvine, 2014.

30

A COMMENTARY ON THE OXFORD PRINCIPLES

Tim Kruger

Origins of the Oxford Principles

The context in which the Oxford Principles (OPs)[1] were drafted was shortly after the Royal Society published its report *Geoengineering the Climate*, just prior to the UNFCCC COP in 2009 at Copenhagen (Royal Society, 2009). With tremendous expectations heaped upon the COP, some commentators were openly hostile to even discussing geoengineering. It was perceived that even raising the hypothetical possibility of such techniques might detract from efforts to cut emissions – the so-called "moral hazard" effect.

The germ of the idea that would become the OPs was a conversation between Steve Rayner and me in November 2009. The UK House of Commons Science and Technology Committee had just issued a call for evidence for their inquiry into "The Regulation of Geoengineering" (HOC, 2010), and we thought it would be useful to submit some draft guidelines for research in this field.

We recognised that the guidelines would be greatly strengthened if experts in a range of fields participated in their drafting. To that effect we invited the involvement of Catherine Redgwell, Julian Savulescu, and Nick Pidgeon – and between us there was expertise in social science, international law, ethics, psychology, and the technical aspects relating to geoengineering. Over November and December 2009, the authors prepared a set of "Draft Principles for the Conduct of Geoengineering Research" (Rayner et al., 2009), which were submitted to the HoC S&T Committee as evidence for their inquiry.

The cross-party committee of MPs would proceed to use the framework of the OPs in their questioning of those who gave oral evidence to the inquiry in early 2010. Their report of March 2010 would state: "While some aspects of the suggested five key principles need further development, they provide a sound foundation for developing future regulation. *We endorse the five key principles to guide geoengineering research*" (House of Commons, 2010, p. 35). The UK Government, in their response to the HOC S&T Committee's report of September 2010, welcomed "the contribution of the Committee and academics in framing the outline of a set of principles to guide geoengineering research" and encouraged their further development (UK Government, 2010).

The HoC S&T Committee's report was published the week prior to the Asilomar Conference on Climate Intervention Technologies (March 2010) that had been organised by the Climate Institute to discuss how to promote the responsible conduct of research on climate engineering. During the conference, Steve Rayner presented the "Draft Principles for the Conduct of Geoengineering Research" to the conference, coining the name "The Oxford Principles".[2]

They subsequently formed the basis of discussions there and the Asilomar Principles for Responsible Conduct of Climate Engineering Research, which emerged from that meeting, acknowledged that they drew from the OPs (Asilomar, 2010, p. 8).

The motivation behind the Principles

The five Oxford Principles are:

i.   Geoengineering to be regulated as a public good
ii. Public participation in geoengineering decision-making
iii. Disclosure of geoengineering research and open publication of results
iv. Independent assessment of impacts
v.   Governance before deployment.

Principle 1: Geoengineering to be regulated as a public good

While the involvement of the private sector in the delivery of a geoengineering technique should not be prohibited, and may indeed be encouraged to ensure that deployment of a suitable technique can be effected in a timely and efficient manner, regulation of such techniques should be undertaken in the public interest by the appropriate bodies at the state and/or international levels.

Commentary

The UNFCCC's preamble acknowledges "that changes in the Earth's climate and its adverse effects are a common concern of humankind" (UNFCCC, 1992). As such there is a desire that activities that might play such a role are not dominated by a small group, be they a subset of the world's governments or powerful business interests. Rather, such activities should be governed in such a way that benefits everyone.

Regulation of geoengineering as a public good should not be read as a rejection of private sector involvement in the development of potential geoengineering techniques – there is an important role that the private sector can play in ensuring that any suitable technique could be deployed in a "timely and efficient manner". A line of discussion in the oral evidence to the HoC S&T Committee raised the concern that this principle was about excluding the private sector – in response, the authors of the OPs submitted a supplementary submission specifically addressing this concern (Kruger et al., 2010).

That response made clear that private sector involvement should be encouraged, albeit with a regulatory framework that would help to stymie the creation of vested interests in this space, and particularly with respect to the issues of patents and other intellectual property rights. It also recognised that with regard to intellectual property issues the heterogeneity of proposed geoengineering techniques meant that a one-size-fits-all approach would not be appropriate. For example, the development of some techniques (biochar) might benefit from normal patent regulations, while for others (stratospheric particles) it may be better to restrict, or even exclude, intellectual property rights.

Principle 2: Public participation in geoengineering decision-making

Wherever possible, those conducting geoengineering research should be required to notify, consult, and ideally obtain the prior informed consent of those affected by the research activities. The identity of affected parties will be dependent on the specific technique which is being researched – for example, a technique which captures carbon dioxide from the air and geologically sequesters it within the territory of a single state will likely require consultation and agreement only at the national or local level, while a technique which involves changing the albedo of the planet by injecting aerosols into the stratosphere will likely require global agreement.

Commentary

In general, two reasons are advanced in favour of public participation. First, a *normative* view – that it is the right thing to do and that a decision can be legitimate only when it has been consented to by those affected. Second, a *substantive* view – that it leads to better decision-making because all information and perspectives can be brought to attention.[3] This Principle, along with Principles 3 and 4, recognises the importance of both views.

The use of the words "prior informed consent" is a deliberate echo of the language used in medical ethics, where respecting the views of the patient is fundamental. But just as in medical situations, while desirable, it may not always be possible to obtain the patient's consent; hence the explanatory text starts with the words "wherever possible". It is obviously important to define in what circumstances such an exclusion might apply.

The extent of public participation in geoengineering decision-making will be determined by both necessity and feasibility. As the explanatory text makes clear, the heterogeneity of geoengineering techniques will result in different requirements for consultation and agreement – techniques without trans-boundary effects could reasonably be determined by involving people in the particular state, while techniques with global effects would require global agreements.

It is also important to note that "public participation" will differ substantially depending on which public is being referred to. As stated in the *Climatic Change* piece: "Differences in political and legal cultures will shape the mode and extent of public participation around the world. Different ideas about democracy and the relationship between individuals and society will engender different understandings of consent" (Rayner et al., 2013). A requirement for "global agreement" as stated in Principle 2 does not mean that universal democracy is a prerequisite to deployment of stratospheric aerosols – what it does mean is that engagement with representatives of countries which may be affected by a technique should be sincere, thorough, and transparent. It may be questioned why such adjectives are needed – after all who would argue for an insincere, slipshod, and opaque dialogue with society? Yet, that is how engagement about new technologies is conducted all too often.

Principle 3: Disclosure of geoengineering research and open publication of results

There should be complete disclosure of research plans and open publication of results in order to facilitate a better understanding of the risks and to reassure the public as to the integrity of the process. It is essential that the results of all research, including negative results, be made publicly available.

Commentary

As previously stated, this Principle, like Principles 2 and 4, is motivated both by normative and substantive perspectives – that transparency is both the right thing to do and is likely to lead to better decision-making. The counterfactual here is that research plans are not disclosed and the results of such research are not published openly. A lack of transparency would undermine trust, as has been seen with other controversial new technologies, such as genetically modified organisms (GMOs) and the nuclear industry.

The motivations for secrecy may be different – for the nuclear industry there are security concerns, while for biotech companies there are commercial reasons – but in both cases secrecy can be used to obscure inconvenient findings and sweep incompetence under the carpet. Indeed, the pharmaceutical industry has been rocked by a series of scandals involving concealment of negative trial results (for example, see Wadman, 2004). Were something similar to happen in the field of geoengineering research, it could devastate public trust and could lead to a backlash against geoengineering researchers and their research.

Moreover, it would be important to know whether any other geoengineering experiments were taking place at the same time – if you wanted to conduct an experiment to test the safety of stratospheric particles as a means to alter the planet's albedo, you would probably want to apply a very low dosage of particles over a long time period (low dosage to minimise any side effects, long duration to allow the signal from the experiment to be discernible from the noise). Were other, unregistered, experiments taking place at the same time the results would not be reliable.

It is thus necessary for research plans to be published – it is hard to see how Principle 2 (public participation in decision-making) could be meaningfully undertaken otherwise. And without open publication of results Principle 4 (independent assessment of impacts) would be severely hampered.

As the authors of the OPs observed in the supplementary memorandum to the HoC S&T Committee about the risks associated with the concealment of negative results: "The highly regarded House of Lords Science and Technology Committee "Science and Society Report" of 2000 concluded that openness and transparency are a fundamental precondition for maintaining public trust and confidence in areas which may raise controversial ethical or risk issues" (Kruger et al., 2010).

The pharmaceutical industry scandals led to the setting up of a national trials registry in the US, as well as similar initiatives elsewhere. It is hoped that a similar register, international in scope, could be set up for geoengineering research pre-emptively – without being impelled by a scandal to do so.

Principle 4: Independent assessment of impacts

An assessment of the impacts of geoengineering research should be conducted by a body independent of those undertaking the research; where techniques are likely to have trans-boundary impact, such assessment should be carried out through the appropriate regional and/or international bodies. Assessments should address both the environmental and socio-economic impacts of research, including mitigating the risks of lock-in to particular technologies or vested interests.

Commentary

This Principle, like Principles 2 and 3, is based on both the normative and substantive advantages of transparency. The validity of any "prior informed consent" hinges crucially on the validity of the information that is used to arrive at a conclusion, and independence of assessment is essential.

A cynic might also identify a more *instrumental* perspective – that engagement processes are carried out because they serve particular interests. It is fair to say that the authors of the OPs support the undertaking of geoengineering research because they believe that research needs to be undertaken – but they do not support the undertaking of geoengineering deployment, which can only be assessed once some research has been undertaken.

As is stated in the preamble of the Principles:

Recognising the fundamental importance of mitigation and adaptation in combating climate change and its adverse effects

Acknowledging nonetheless that if, in the near future, the international community has failed to reduce greenhouse gas emissions and urgent action is needed to prevent catastrophic climate change then it may be necessary to resort to techniques involving deliberate large-scale intervention in the Earth's climate system ("geoengineering")

Ensuring that, in the event such resort is necessary, potential geoengineering techniques have been thoroughly investigated to determine, which, if any, techniques will be effective in addressing the issue of climate change without producing unacceptable environmental and socio-economic impacts

Stressing that research into geoengineering techniques does not lead inevitably to deployment and that principles to govern research may need to be adapted to guide decisions regarding deployment if any.

Transparency is thus a necessary – though not sufficient – requirement to obtain the "social licence to operate" that other novel technologies such as nuclear power and GMOs have to a significant extent struggled to obtain. But public participation is not just a "box-ticking exercise", or a means to the end of obtaining the necessary social licence. I reject such a rationale as cynical, though I do not doubt that there will be some who might embrace transparency for that purpose. But for me (and those I choose to associate with) transparency is not a mere "hygiene factor", but has both moral and practical value.

Principle 5: Governance before deployment

Any decisions with respect to deployment should only be taken with robust governance structures already in place, using existing rules and institutions wherever possible.

Commentary

This Principle reflects the view that governance should precede deployment. The question this raises is: should governance precede experimentation? At what point does experimentation raise the same, or sufficiently similar issues, as full-scale deployment? It is not clear where the dividing line is between small-scale experiments – which would have no material impact on the environment – and large-scale experimentation verging on small-scale deployment. Yet, it is important that we establish the location of this Rubicon before we inadvertently cross it.

The role of the Oxford Principles in the SPICE project

A practical use of the OPs was in the decision to institute a stage-gate into the funding of one part of the SPICE project. SPICE (Stratospheric Particle Injection for Climate Engineering) is a project that received funding in March 2010 from two UK research councils – EPSRC and NERC – through the "Geoengineering Sandpit" (a funding mechanism designed to stimulate cross-disciplinary research in emerging fields). The project comprises three elements, of which the first two were relatively uncontroversial and involved computer modelling and laboratory-based research. The third element, the so-called "test bed", intended to deploy a 1km hose suspended from a balloon in order to test a transportation mechanism for materials capable of enhancing the planet's albedo in the stratosphere.

That the proposed test bed experiment would have been environmentally benign is self-evident – it involved transporting a bathtub load of water to an altitude well below that of the stratosphere. But the recognition that this would be the first such experiment in the UK to research a geoengineering technique in the open environment created the need to engage in a dialogue with stakehold-ers – the experiment may have had all the necessary legal licences to operate, but did it have a social licence?

As one of the mentors (decision-makers) at the Sandpit, I made the sugges-tion that the decision as to whether to fund the test bed element of the SPICE project should be stage-gated – that the funding be ring-fenced for the use of the project, but released only on the satisfactory completion of the stage-gate. The suggestion was adopted by the other mentors and integrated into the proposal – funding was agreed unconditionally for the first two elements of the project and agreed conditional on the successful completion of the stage-gate for the test bed element. The exact design of the stage-gate was not agreed during the Sandpit but later determined by a separate body with the expertise required. Yet, fol-lowing a rocky process, the test bed experiment was first delayed, then eventu-ally cancelled.

The reasons for this cancellation are the subject of some debate. However, two key reasons have been given: the engagement process with stakeholders with an interest in this field was not completed satisfactorily, and a dispute about insuf-ficient disclosure of intellectual property held by a participant and a mentor at the Sandpit. My contention is that the former matter was resolvable – the engage-ment process could have been extended and improved, while the latter matter was fundamentally irresolvable. Some people have characterised the intellectual property issue as the straw that broke the camel's back – my view is that it was more of a steel girder than a straw – it was sufficient to break the camel's back all on its own.

It is fair to say that when the decision to stage-gate the funding of the test bed was made, it was not popular with those managing either of the two projects that were impacted by it – the SPICE project itself saw it as a bureaucratic bur-den, and the IAGP (Integrated Assessment of Geoengineering Proposals) project believed that an accelerated public engagement process had been pushed upon it. However, people involved in both projects recognise the value that such a stage-gating process has had in making researchers reflect more deeply on the consequences of their research and the need to obtain a social licence to operate for controversial research. It suffices to say, at this point, that the principle of engagement with those affected by geoengineering research has been established, and we are learning lessons as to the complexity of such a process.

Future developments

So what next for the governance of geoengineering research? There are many opinions as to how such research should be governed, including opinions that no research should currently be undertaken outside of a laboratory setting (Robock, 2012). Nevertheless, the authors of the OPs believe that there is value in seeking to operationalise the OPs – transforming the broad principles into detailed guidelines that are implementable.

There has been much discussion about producing a research register, which draws specifically from Principle 3 on the disclosure of geoengineering research and open publication of results. It could also incorporate the dissemination of the independent assessment of impacts detailed in Principle 4 and could inform Principle 2 on public participation in decision-making, as it is hard to see how meaningful consent can be obtained without the information required to make such decisions. Work is ongoing on designing a research register, with debate centring specifically on forms of research that should and should not be included in such a register.

Work on Principle 5 (governance before deployment) is proceeding slowly. Proposals have been put forward to use the framework developed by the London Convention/London Protocol for the governance of ocean fertilisation research as the basis for governance of other geoengineering research (Markus and Ginzky, 2011). Work is ongoing in assessing whether text incorporated in other existing treaties governing the management of commons (for example the Antarctic Treaty) could be adapted for use in governing geoengineering research.

It is important to distinguish both between the potential impacts of the wide range of proposed techniques and also between the scale at which research may be undertaken. There is not going to be a one-size-fits-all approach that is valid for all proposed techniques or for all scales of experimentation or deployment. To this end, the recent paper in *Climatic Change* on the Oxford Principles recommends the "development of research protocols for each stage of development of the technology … to be interrogated by a competent third party as part of a stage-gate process" (Rayner et al., 2013). As described earlier, such a stage-gate process was used in the SPICE project. In that instance, it was a novel approach and thus was implemented in an *ad hoc* manner. My hope is that such a stage-gate process could become the norm and technology-specific research protocols and public engagement procedures could be developed appropriate for the different technologies and scales of implementation.

The authors of the Principles have reiterated that the OPs are draft principles and the commentary attached to them in the submission specifically invites others to develop them further (ibid.). It would be strange indeed if we did not invite a broader involvement in the development of such guidelines – the recognition that the subject of geoengineering deserves and requires broad engagement is the motivation that lies behind all the Oxford Principles.

First published online as an opinion article in 2013.

Acknowledgements

This work was supported by the Oxford Geoengineering Programme, Oxford Martin School, University of Oxford. I would like to thank Steve Rayner and Clare Heyward for their comments and advice relating to earlier versions of this piece.

Notes

1 A thorough and detailed analysis of the Oxford Principles (OPs) has been published elsewhere (Rayner et al., 2013). This piece is a personal reflection on the origins of the OPs, the motivation behind each of the principles, their role in the SPICE project, and how they might be developed in the future.
2 The OPs were named for the location of the meeting at which they were initially drafted, rather than because of the affiliation of the authors – Catherine Redgwell was at University College London and Nick Pidgeon was at Cardiff University.
3 The analysis of the normative, substantive, and instrumental motivations for public participation detailed here and on subsequent pages draws on Andy Stirling's work on the social appraisal of technology (Stirling, 2008).

31

ENVIRONMENTAL INSTITUTIONS, INTERNATIONAL RESEARCH PROGRAMMES, AND LESSONS FOR GEOENGINEERING RESEARCH

Arunabha Ghosh

The existing landscape of multilateral environmental agreements varies in terms of their relevance to governing (largely, prohibiting) the deployment of geoengineering technologies. There is, however, a governance gap regarding R&D activities on geoengineering. No existing institution appears to have the mandate or capacity to govern the upstream process of laying down proactive research and governance mechanisms. Meanwhile, research activities are gaining momentum, even though the vast majority of researchers might currently be concentrated in a few developed countries, thus raising questions about the legitimacy of the research and exposing governance deficits. What lessons can be drawn from other international research endeavours to design coordinated scientific research in solar geoengineering?

Attempts to coordinate geoengineering research internationally hinge on technical and scientific demands on one hand, and ethical and political considerations on the other. Some researchers argue that prohibitions on geoengineering research violate the basic principle of freedom of science. Others contend that, if the research has cross-border dimensions or risks, then it would have to be governed in some form, although governance need not mean only prohibition or formal international treaties. International cooperation over governing research is not a given and would depend on the mix of material interests and ethical concerns for research partners as well as those countries outside the scope of research programmes.

This chapter argues that, given the nature of research, funding requirements, political imperatives, and the need to win informed public acceptance, internationally coordinated geoengineering research programmes would be necessary. The chapter also draws on examples from past international research programmes (World Climate Research Programme, the European Organization for Nuclear Research, and the International Thermonuclear Experimental Reactor, among others) to argue that several key characteristics define successful research endeavours: inclusiveness, transparency and review, public engagement, and precaution. Finally, the chapter discusses operational aspects of international research programmes, namely research capacity, flexible funding, establishing liability, and intellectual property.

This chapter considers the question largely in the context of research connected with solar radiation management (SRM). Although the text uses geoengineering and SRM interchangeably, it does not imply that all the findings and recommendations found relevant to SRM research would automatically apply to carbon dioxide removal (CDR) methods.

International governance gap for geoengineering research

Although geoengineering research is being discussed at the national legislative level (note the congressional and parliamentary hearings in the United States and United Kingdom (HOC, 2010; Gordon, 2010; USGAO, 2010a), respectively, and government reports on geoengineering in Germany (Rickels et al., 2011)), the potential cross-border environmental externalities associated with solar geoengineering mean that some form of international governance arrangement would be certainly demanded, if not inevitable.

But almost no international agreements or decisions yet exist that are specific to geoengineering, with the exception of the broad decision by the Convention on Biological Diversity (CBD) at Nagoya. However, under certain interpretations, the rules and mandates of several international organisations and multilateral environmental agreements (MEAs) have potential intersections with geoengineering. A non-exhaustive list might include the CBD; the UN Framework Convention on Climate Change (UNFCCC); the UN Environment Programme (UNEP); the UN Convention on the Prohibition of Military or Any Other Hostile Use of Environmental Modification Techniques (ENMOD); the Antarctic Treaty System; the Convention on Long Range Transboundary Air Pollution (CLRTAP); the International Maritime Organization (IMO); the Montreal Protocol on Substances that Deplete the Ozone Layer; and the UN Outer Space Treaties (Blackstock and Ghosh, 2011).

An integrated governance framework should, ideally, cover all stages of SRM technology development: research (i.e. computer modelling, laboratory activities), field testing and deployment. Yet, the current landscape is disposed to what can be called *downstream* governance (constraining field testing or full deployment) rather than *upstream* governance (coherence of initial research principles, review of research outputs, and building forums to scope and frame an emerging issue comprehensively). Downstream institutions can govern deployment and its repercussions by either ruling on violations of allowable emissions or social or physical impacts, or by establishing liability after deployment, in part by determining the motivation for such action (e.g. ENMOD). Significantly, many institutions appear more aligned with simply prohibiting – or at least severely limiting the use of – geoengineering. No existing institution has yet developed a comprehensive assessment process for SRM or laid down proactive research and governance mechanisms.

Does SRM necessarily require international governance?

If governance arrangements were to develop, they would need to account for the trans-boundary nature of research, field tests, and deployment – and their impacts. The principles affecting the governance of geoengineering will depend on how it is categorised (Morgan and Ricke, 2010; Morgan et al., 2013). Irrespective of the scale of activity, there are international dimensions of geoengineering research.

Computer modelling

Virtually all solar geoengineering research uses computer models. Whether testing ideas of sulphate injections in stratospheric clouds or brightening marine low clouds, their effects are calculated with the same computer models that are also used to study the climate system.

SRM research would have to build on existing international collaborations in climate science. For instance, in preparation for the IPCC AR5 report, the Climate Model Intercomparison Project 5 (CMIP5) was conducted by about 20 general circulation model research groups around the world (Taylor, Stouffer, and Meehl, 2008). The Geoengineering Model Intercomparison Project (GeoMIP) has piggybacked onto that experiment. GeoMIP examines how reducing solar radiation would reverse warming from CO_2 in many of the CMIP5 and CMIP 6 computer simulations, ensuring comparability of SRM modelling results for the first time across a number of climate models (Kravitz et al., 2011; 2013; 2015). This experiment was endorsed by the World Climate Research Program's Working Group on Coupled Modelling (WGCM) as a "Coordinated CMIP Experiment".

Field experiments

The environmental risk of field experiments would vary by scale but who is to decide what is a small- versus a medium- or large-scale experiment? With increase in scale, each experiment would have to be separately reviewed and approved. At least three scenarios should be considered. The first is when the experiment is entirely privately funded. While such activity could fall outside the purview of national governments (depending on the scope of domestic laws), its international consequences would still demand attention. If the scale of the experiment is expected to have trans-boundary consequences, then appropriate international governance mechanisms would be demanded. What kind of obligations do private research institutions or consortia have towards the rest of the world? If national laws are ambiguous, would laws emanating from regional or multilateral institutions be sufficient to regulate such activity?

A second scenario arises when a small number of countries decide to collaborate on a field experiment. Here, too, the scope of the research collaboration would be determined by the countries concerned. They might or might not choose to allow other countries to join the research group. There are also other concerns about the transparency of the research, whether the data would be available to non-members of the research group. The most important question would be whether international laws and organisations could have any jurisdiction over a subset of countries that have voluntarily chosen to come together in a research project. If the answer is unclear, then the opposition to field experiments would also intensify.

A third scenario is a multi-country project. Here, a large number of countries could decide to engage in experiments of a specified scale, with each country contributing to the costs or scientific resources or both. Alternatively, the experiment could commence with fewer countries but with provisions to include others. The parameters for admission could vary, as could the basis for joining the project (a formal treaty or a loose collaboration).

Deployment

At the other extreme, one or more countries may decide to deploy geoengineering technologies in future (these technologies do not exist at present). An engineering analysis has suggested that a small fleet of high-altitude aircraft could succeed in conducting experiments on a large scale (McClellan et al., 2010). So far, however, there are no *in situ* geoengineering experiments that are being conducted.[1]

Actors, scales, motivations, and governance

The above discussion suggests that, in the absence of a governance framework, how the scales of geoengineering activities are interpreted will depend on the actors (private or public) promoting these activities. Some scholars argue that, since actors have varied interests, the definitions of scale – whether measured in time, space, or emission amount beyond which environmental impacts might be possible – need to be resolved in advance (Robock et al., 2010).

But who would make those decisions about scale and impacts? Note that governance of geoengineering does not necessarily mean an international treaty. But any governance arrangement would have at least three essential functions: making decisions, monitoring actions, and resolving disputes (Chayes and Chayes, 1995; Abbott and Snidal, 2009; Ghosh and Woods, 2009; Ghosh, 2010, 2011b). These functions could be undertaken by groups of scientists, by apex scientific bodies, by national regulation, by non-binding international principles, or by bilateral, mini-lateral, or multilateral treaty arrangements. The design of the governance arrangements would be a function of interest-based concerns that various actors have and the balance of ethical concerns.

Material interests in SRM stem from the current levels of uncertainty with regard to the science. Some countries and scientists might argue that there is a need to retain the freedom to experiment to improve our knowledge to make informed decisions (Crutzen, 2006; Novim, 2009; Morgan and Ricke, 2010; Benedick, 2011). But rules are also needed to rein in runaway unilateral action in an uncertain technological field (Victor, 2008; ETC Group, 2010b). In a sense, countries are concerned about unanticipated outcomes.[2] Moreover, countries may wish to rein in unilateral action fearing others might gain a technological edge without clarity about their intentions. In short, actors might favour rules that give them maximum flexibility while keeping others off balance.

Ethical concerns, in terms of the legitimacy of a governance structure, translate into processes and outcomes. On the process side, actors wish to participate in forums at which rules are drafted, they have the power to influence such rules, and they can be fully informed before giving consent to governance arrangements. Countries, civil society, and scientific communities also have concerns about outcomes, such as if research capability influences governance arrangements, how growing capability could shift intents about geoengineering, and how actions would be monitored or disputes resolved.[3]

Table 31.1 illustrates, schematically, how the interaction of material interests and ethical concerns would influence how decisions are made, actions monitored, and disputes resolved (see detailed scenarios in Ghosh, 2011a). If actors' interests are to retain maximum flexibility for their research agendas, the scope of governance will be limited and any adjudication will be decentralised. If the intention is to constrain others, more formal rules will be drawn up, with third-party reporting and adjudication by higher authorities. Process legitimacy will depend on how inclusive procedures are for decision-making, review, and dispute settlement, while outcome legitimacy will depend on how voting rights are determined, the quality of reporting, and the ability to enforce decisions.

Such governance arrangements are not only for inter-country forums. The same factors would also influence the governance of geoengineering research or field experiments even among groups of scientists or between different research groups, private sector entities, and civil society organisations. It is not necessary that all forms of geoengineering require inter-state governance. But if research were occurring among scientists in different countries, or if field tests were expected to have cross-border impacts, then irrespective of the scale of activity, governance arrangements, broad or narrow, would have an international dimension.

Could governance arrangements become excessively prohibitive at early stages of research, or overly permissive with regard to field tests and deployment? These are certainly possibilities, but they are contingent on the balance of

TABLE 31.1 Which functions for what motivations?

| | Interest-based concerns | | Ethical concerns | |
	Maintain flexibility	Constrain others	Process legitimacy	Outcome legitimacy
Making decisions	Scope of governance limited	Scope of governance broad	Inclusive process vs. ease of decision-making in small groups	Equally weighted voting rules vs. capability-driven voting
Monitoring actions	Self-reporting	Institutional reporting plus verification	Inclusiveness of review procedures	Quality and timeliness of reporting
Resolving disputes	Decentralised adjudication, including market instruments	Centralised adjudication plus centralised/decentralised enforcement	Ease of access to dispute settlement forums	Ability to enforce decisions against powerful countries

Source: Adapted from Ghosh, 2011a.

power among actors seeking to design SRM governance. The likelihood of such outcomes should not become a reason for avoiding governance altogether. As argued earlier, the demand for governance arises from how SRM research, field tests, and deployment are conceptualised and by the actors that drive them as well as the actors that are not included. If the extremes of uninhibited unilateral action and outright prohibition of early-stage research have to be avoided, then geoengineering research would have to be coordinated internationally. What principles would such research programmes follow and what lessons can they draw from other international research endeavours?

Lessons from other international research programmes

Some aspects of SRM research cannot be conducted solely within national borders. The nature of the scientific inquiry (such as measuring ocean acidity, carbon dioxide concentrations in the atmosphere, and the impact on monsoons and soil moisture) requires international research programmes (Crutzen, 2006; Caldeira and Wood, 2008; Novim, 2009). There are also financial constraints for individual countries. There are demands for being inclusive in the research process. There are political constraints about who contributes and who controls the research activity. And there are issues about public engagement in the research activity.

Although SRM research is controversial and replete with uncertainties, there are several examples that could offer lessons on how international research collaborations originate, how they are funded and governed, and how they expand their membership. Admittedly, the examples in this section are not a comprehensive list of international scientific research programmes. But their selection is not arbitrary. They draw on other examples of research, which required international modelling efforts (IGY, WCRP, HUGO), or activities with large financial contributions from many countries thanks to the infrastructure required (CERN, ITER), have included developing countries actively (CGIAR, ITER), or been politically astute in bridging divides (ITER) or permitting in kind contributions (ITER), or have had public engagement due to the potential for adverse cross-border or regional consequences (nuclear waste management).

International Geophysical Year

The International Geophysical Year (IGY), lasting from July 1, 1957 to December 31, 1958, was the world's first sustained multinational research collaboration on the environment. The International Council for Science (ICSU), an independent federation of scientific unions, took the lead in organising and funding the IGY. A Special Committee for the IGY (CSAGI) served as the governing body. Representatives of 46 countries originally agreed to participate in the IGY; by its close, 67 countries had become involved.

World Climate Research Program

The World Climate Research Program (WCRP), established in 1980, was jointly sponsored by ICSU and the World Meteorological Organization (WMO). It has also received support from UNESCO's Intergovernmental Oceanographic Commission (IOC) since 1993. Aiming to improve scientific understanding of the Earth's physical climate system, WCRP studies the global atmosphere, oceans, sea ice, land ice, and the land surface. The three sponsoring organisations have appointed, by mutual consensus, a Joint Scientific Committee comprising 18 scientists. The research is itself conducted by scientists in national and regional institutions, laboratories, and universities. WCRP regularly informs the UN Framework Convention on Climate Change and its subsidiary bodies. Peer-reviewed publications by scientists affiliated to the WCRP underpins much of the work of the Intergovernmental Panel on Climate Change.

European Organization for Nuclear Research

The European Organization for Nuclear Research (CERN), established in 1954, is the world's largest particle physics laboratory, situated on the Franco–Swiss border. Run by 20 European countries,[4] the CERN Council has two representatives from each member state, one representing the government and the other her/his country's scientific community. Decisions are by simple majority and based on one-country-one-vote, although the Council usually aims for consensus (CERN, *The Structure of CERN*). CERN spends much of its budget on building new machines (such as the Large Hadron Collider) and only partially contributes to the cost of the experiments. Other countries and organisations have observer status (the European Commission, India, Israel, Japan, Russia, Turkey, UNESCO, and the United States) and 57 other countries have cooperation agreements or scientific contacts with CERN (CERN, *Member States*). Consequently, scientists from more than 600 institutes and universities around the world use CERN's facilities.

International Thermonuclear Experimental Reactor

The International Thermonuclear Experimental Reactor (ITER) is an international research and engineering project, which is currently building the world's largest and most advanced experimental nuclear fusion reactor. ITER originated from discussions in 1985 when President Gorbachev, following discussions with President Mitterrand, proposed to President Reagan that an international project be set up to develop fusion energy for peaceful purposes. ITER began as a collaboration between the European Union, Japan, the former Soviet Union, and the United States (ITER, *The ITER Story*, n.d.). Its current members are the European Union (contributing 45–50 per cent of the cost) and China, India, Japan, South Korea, Russia, and the United States, each contributing 9–10 per cent (ITER, *ITER: Our Contribution*, n.d.). Originally expected to cost around €5 billion, the estimates are now in the region of €10–15 billion, with growing pressure for more transparency about the costs of the project (Amos, 2010). The process of selecting a site for the ITER project ran from 2001 to 2005 culminating in the choice for Cadarache, France. Since Japan lost out on its proposed site, it was promised 20 per cent of research staff (in return for only 10 per cent of the funding) as well as the right to propose the director general. Further, another research facility for the ITER project would be built in Japan, for which the European Union has agreed to contribute about 50 per cent of the costs (ITER, *The ITER Story*, n.d.).

Nuclear waste management[5]

Nuclear waste management and disposal have also benefited from international collaboration. As with geoengineering, these topics raise complex questions about technology, Earth science, long-term stewardship, and public engagement. A number of inter-country collaborations, notably with the Swedish nuclear waste programme, allowed the international community to share the burden of technology development and formulate technical norms for characterising and analysing the behaviour of nuclear waste repository sites. What started as a national programme of waste management in Sweden resulted in, first, a collaboration with Finland, which then became the basis of a European Technology Platform.

Much of this collaborative technical work was used in Sweden and other countries (though not in the United States) as a basis for licensing facilities and for securing public acceptance of individual countries' nuclear waste management plans. An EU-wide nuclear waste storage facility is now being considered under the Strategic Action Plan for Implementation of European Regional Repositories (Stage 2) (SAPIERR II); this too has strong support from Sweden (European Commission, 2008). Countries that participated in these research programmes provided funding, agreed on research goals and established a formal process for adaptive management, which allowed the programme to take credit for the results it achieved.

Consultative Group on International Agricultural Research

The Consultative Group on International Agricultural Research (CGIAR) was established in 1971 although its roots lay in agricultural research that began in Mexico in 1943 (funded by the Rockefeller Foundation). By the 1960s, the International Maize and Wheat Improvement Center (CIMMYT) in Mexico and the International Rice Research Institute (IRRI) in the Philippines had been established. The CGIAR was created to coordinate agricultural research and food security measures that were being employed in several developing countries, thus evolving a network of research institutions, which also included the International Centre for Tropical Agriculture (CIAT) in Colombia and the

International Institute for Tropical Agriculture (IITA) in Nigeria. Currently, 15 centres are supported in the CGIAR network (CGIAR, n.d.). The CGIAR, in turn, received funding support from the Food and Agricultural Organization, the United Nations Development Programme, and the World Bank during its inception.

In December 2009, a new institutional model was adopted to offer programmatic support via a new CGIAR Fund, to bring balance to governance by including donors and researchers, and to create a new legal entity that would bring the research centres together. The Fund's governing council now includes eight donor representatives, eight representatives from developing countries and regional organisations, and six representatives from multilateral organisations and private foundations. An Independent Science and Partnership Council was also established to provide advice and expertise.

Human Genome Project

The Human Genome Project (HGP) was a 13-year-long international research programme aimed at discovering the 20,000–25,000 human genes and to complete the sequence of the 3 billion chemical base pairs that constitute human DNA. The project, which ran during 1990–2003, was funded by the US Department of Energy and the US National Institutes of Health National Human Genome Research Institute. They together spent nearly $3.8 billion on the project (Human Genome Project, *Human Genome Project Budget*, n.d.). At least 18 countries also established research programmes, including China, France, Germany, Japan, and the United Kingdom. The Human Genome Organisation (HUGO), conceived in 1988, helped to coordinate some of the international research effort (Human Genome Organisation, n.d.).

Lessons for SRM from other international research programmes

The purpose of the above examples is not to draw a like-for-like comparison with SRM research. Surely, there would be differences, in the complexity of the science, costs of research, testing and deployment, extent of trans-boundary risks, and so forth. But these examples highlight certain lessons, which might be relevant to organising international research programmes for SRM as well. They are also similar to voluntary principles adopted for geoengineering research, such as the Oxford Principles (Rayner et al., 2009; Rayner et al., 2013) and the Asilomar Principles (Asilomar Scientific Organizing Committee, 2010).

Inclusiveness

Inclusion may be promoted through both voluntary and treaty-based participation. Membership during the IGY was partly voluntary and partly based on international treaties (such as the Antarctic Treaty). The 2006 ITER agreement established an international organisation responsible for all aspects of the project: licensing, hardware procurements, construction, the 20-year operation period, and the decommissioning of ITER at the end of its lifetime (*The ITER Story*, n.d.). For SRM research, individual scientists could be seconded to collaborate on projects in other countries, thereby helping to build an international network of researchers rather than drive nationally determined projects.

Transparency and review

Any credible scientific research programme must undergo rigorous peer review. But the research must also be conducted in a transparent manner, especially if it has cross-border risks. One of the HGP's goals was to store the DNA sequence in publicly available databases, which are housed in GenBank, a public database operated by the US National Center for Biotechnology Information. All the major research papers published during the project were given free access by *Nature* and *Science* journals. Currently, for solar geoengineering research, transparency is mainly gained from numerous conference sessions and workshops held each year. The IPCC held an expert meeting on geoengineering in June 2011, to discuss both the state of geoengineering science, as well as state of the art research in the economic, ethical, political, and legal dimensions of geoengineering. Geoengineering is now also addressed in the IPCC's Fifth Assessment Report (IPCC, 2013). However, if most SRM research occurs in a few developed countries, and most related academic conferences are also held there, some might question their legitimacy.

As Table 31.1 showed, process and outcome legitimacy of monitoring depends on how inclusive the review procedures are and how timely and relevant is the information provided. Options include institutionalised self-reporting (research consortia report periodically on activities), to cover scientific methods and results, financing sources, governance mechanisms, and so forth. Transparency is further strengthened by independent review procedures, say by the governing council of a research consortium, or by third parties (as occurs under the WTO's Trade Policy Review Mechanism or the IMF's Country Reviews).

Public engagement

Further, it is not sufficient to publish results in refereed journals. For instance, ICSU's Principle of the Universality of Science, laid out in Statute 5, demands equitable access to data, information, and research materials. But unless the data is presented in accessible, usable, and comparable formats, it would be difficult for governments and research institutions to engage other sections of society and inform them about the potential impacts (domestic or trans-boundary) of geoengineering experiments. As a result, numerous public engagements have been held across the world, but they would have to intensify to widen the reach of the debates around geoengineering (see, for example, Chapter 23 of this volume).[6]

The absence of such engagement could backfire on research activities if public opposition results in all initiatives being banned, irrespective of the scale and nature of the research.

Precaution

Even though geoengineering research might be necessary to prepare for a "Plan B" against the risk of severe climate impacts, it is also important that all caution be exercised in the scope and scale of such research. Precaution would imply that high-risk technologies are avoided entirely, or a moratorium is agreed against their deployment, at least until appropriate governance mechanisms for establishing liability are established. The calculation of risk itself would be contingent on factoring in the uncertainties and ignorance (technical, political, and social) associated with emerging research and technologies (Long and Winickoff, 2010).

Operational aspects of designing international research programmes

The examples of antecedent international research programmes also offer insights for operational aspects of coordinating SRM research across multiple institutions and jurisdictions.

Research capacity

For a broad-based research agenda to develop, capacity is a key consideration. Geoengineering research activities will have to devote greater attention to emerging economies and poorer countries, by starting to identify potential institutions that could be drawn into a network of international research collaborations. Efforts would be needed (combined with financial support) to engage these institutions and build local research capacity, say by developing segments of projects focused on measuring the applicability and impact of the technology in local conditions. Another approach would be to source inputs from developing countries to build components of larger infrastructure, as is planned for ITER. Again, at CERN developing countries have been asked to produce materials that are used to build particle detectors (Harvard Model UN, 2011). Research should also draw upon local experience to understand the social and political dimensions of geoengineering. For instance, the HGP's stated goals include studying ethical, legal, and social issues that were expected to arise from the project (Human Genome Project, *HGP Research Area*, n.d.).

Flexible funding

One major problem with promoting international geoengineering research is raising and monitoring funds. In continental Europe and the United Kingdom, calls for proposals for research on geoengineering have emerged recently. In the United States, by contrast, such funding comes through the normal funding process; there is no national research programme. One option is to consider funding "in kind" whereby member institutions or countries are allowed to offer staff capacity, institutional resources, or material inputs as ways to participate in a joint project (as ITER permits). The transparency of funding channels and the openness of the intellectual property regimes vis-á-vis geoengineering research would be important to ensure that such efforts are not rewarded by exclusive patents.

Responsibility and liability

With many parties involved in research, responsibility for anticipated and unanticipated adverse outcomes has to be ascribed and limits on liability established. Where international scientific research has created independent institutions, liability clauses are more explicit than where loose groupings of scientists engage in collaborative research. CERN assumes the expense of insuring against risks of "fire, explosion, natural disaster and water damage" for all items belonging to both the collaboration and collaborating institution, once they have been delivered to the CERN site. CERN also insures members of collaborating institutions from third-party liabilities incurred at CERN during an experiment. However, such liability is limited and there is no warranty that it would be sufficient to cover for the full extent of the risks involved (Sections 5.4, 5.5, and 5.6 of European Organization for Nuclear Research, 2008). Similarly, Article 15 of the Agreement of the ITER Organization provides for contractual as well as non-contractual liability assumed by the organisation. The European initiative for Implementing Geological Disposal of Radioactive Waste Technology Platform (IGD-TP) offers the option of deciding on governance questions through a legal agreement or by setting terms of reference for joint activities. For a legal agreement, every organisational partner has to agree even if the joint activity is among a subset of all members (see management guidelines in IGD-TP, 2012).

Intellectual property and access to data

Intellectual property rights in international research programmes are controversial because each country has different rules and there are ethical questions about whether research conducted in the public interest should be commercialised on a private basis.

The extent of public and private commercial interest varies. The IGY's organising committee was categorical that "all observational data shall be available to scientists and scientific institutions in all countries" (NOAA, n.d.). Under the HGP, results were available on open source platforms. Research involved public collaborations as well as a private firm, Celera Genomics. When the latter filed preliminary patent applications on 6500 whole or partial genes, thus threatening the free flow of data, the University of California-Santa Cruz published the first draft of the human genome. The "Bermuda Principles", by which data are expected to be released within 24 hours, have been used to counteract the normal practice of making experimental data available only *after* publication – and the entire gene sequence was freely available.

CERN normally retains ownership of technologies that it develops or concludes joint ownership and exploitation (commercial and free use) agreements with other parties.[7] CERN also follows the "open science" model, whereby methods and results are disclosed. Revenues from commercialisation are divided among those who developed the technology, the related CERN department, as well as a special fund to support technology transfer. If there is a conflict between revenue generation and dissemination, dissemination takes precedence (WIPO, 2010a). In August 2010 CERN signed a deal with the World Intellectual Property Organization (WIPO) to facilitate technology transfer (WIPO, 2010b).

ITER allows a member state to acquire rights to IP that it has generated but retains rights over property created by the ITER Organization or its staff (Articles 4.1.1, 5.1.1 and 6.1 in Parties to the Establishment of the ITER, 2006). IP created jointly by members and the ITER Organization are co-owned. There are clear rights of royalty-free access for other members (Articles 4.1.2 and Article 5.1.2), rights of sub-licensing for use by third parties (Articles 4.1.3 and 5.1.3), and even for licensing to third parties of non-members (Articles 4.3 and 5.3).

These examples suggest that the results of scientific research conducted in the public interest are expected to be widely shared. Since climate change threatens humanity and the impacts of geoengineering are expected to be of trans-boundary scale, the case for publicly available data is strengthened. Since solar geoengineering entails risks, it is imperative that any research is treated as affecting the general public interest (Rayner et al., 2009). Government-funded research should, therefore, remain in the public domain, while privately funded work should have limits on proprietary knowledge.

Conclusion

This paper has focused on the governance of geoengineering research. It showed that although several multilateral environmental treaties might have some relevance to geoengineering, there is a governance gap when it comes to research. Depending on the scale and scope of research, field testing, and deployment, there are several aspects that could benefit from internationally coordinated efforts. Numerous past and ongoing international research programmes suggest that there are some basic principles that are key to successful endeavours: inclusiveness, transparency and review, public engagement, and applying the precautionary principle. The pursuit of the principles of open scientific collaboration does not mean that only one kind of institutional design is possible. Operationalising an international research programme means that the interested parties would have to account for variance in research capacity, develop flexible funding mechanisms, outline clear liability rules, and decide on ownership of and ease of access to intellectual property. The challenge with geoengineering research on an international scale is not merely the coordination of the efforts, but developing cooperative mechanisms that reduce uncertainties, increase trust, and are legitimate in the eyes of people and countries that are left outside the process.

First published online as a working paper in 2014.

Notes

1 A Russian experiment, which used tropospheric aerosols, had been mistakenly labelled geoengineering. See Izrael et al., 2009.
2 Uncertainties about rainfall and the hydrological cycle (Bala et al., 2008; Brovkin et al., 2009), tropical forests (Eliseev et al., 2010), ozone layer (Royal Society, 2009; Heckendorn et al., 2009), oceans (Scott, 2005; Trick et al., 2010), and the so-called "termination effect" (also see Robock, 2008; Robock et al., 2008; Robock et al., 2009; Leinen, 2011).
3 Ethical concerns stem from worries about moral hazard (Caldeira and Wood, 2008; Keith et al., 2010), ascertaining intent (Fleming, 2007; Barrett, 2008), cross-border impacts (ETC, 2010a; Banerjee, 2011; NGOs letter, 2011), and intergenerational equity (Burns, 2011; Brown Weiss, 1992; UNFCCC Art. 3(1)).
4 Austria, Belgium, Bulgaria, the Czech Republic, Denmark, Finland, France, Germany, Greece, Hungary, Italy, the Netherlands, Norway, Poland, Portugal, the Slovak Republic, Spain, Sweden, Switzerland, and the United Kingdom.
5 This discussion partly draws on an email communication with Jane Long.
6 Workshops have been conducted in Singapore (July 2011), New Delhi and Tianjin (September 2011), Ottawa (January 2012), Africa (2012 and 2013), among other places. Climate Frontlines (a "forum for indigenous people, small islands and vulnerable communities"), in collaboration with the Convention on Biological Diversity, has opened an online discussion on geoengineering: http://www.climatefrontlines.org/en-GB/node/
7 CERN's policy for IP management in technology transfer is available at: http://kt.cern/technology-transfer/ip-management/cern-ip-policy

32

THE SOLAR RADIATION MANAGEMENT GOVERNANCE INITIATIVE

Advancing the international governance of geoengineering research

Alex Hanafi and Steven P. Hamburg

The issues associated with geoengineering research and deployment extend far beyond the science of how to safely use the technology. The most difficult issues lie in the areas of ethics, politics, and governance. As communities and policy-makers around the world face the risks presented by a rapidly changing climate, understanding the scientific, ethical, and governance issues at the core of geoengineering research will be critical to making informed decisions about response options. Recognising these needs, the Royal Society, Environmental Defense Fund (EDF), and TWAS (the academy of sciences for the developing world) launched the Solar Radiation Management Governance Initiative (SRMGI) in 2010. As an early programme designed to build capacity for understanding, cooperation and practical action on SRM research governance, SRMGI offers preliminary lessons for emerging regimes of geoengineering governance.

Solar radiation management (SRM, also known as solar geoengineering) is a set of theoretical proposals for cooling the Earth by reflecting a small amount of inbound solar energy back into space. The potential importance of SRM should not be underestimated, as it may be seen as the only way to address rising global temperatures quickly, should it be considered necessary.

SRM does not provide an alternative to reducing GHG emissions. The overall effects of SRM for regional and global weather patterns across the globe are likely to benefit some more than others, and intervention will always involve uncertainty and incur risks. The most that could be expected from SRM would be to serve as a short-term tool to manage some climate risks and buy time for more permanent solutions to climate change, if efforts to reduce global greenhouse gas emissions prove too slow to prevent atmospheric greenhouse gas concentrations from causing severe suffering due to disruption of the Earth's climate.

At this point, we don't yet know enough to understand the full implications of intentional climate modification and whether or not, or under what future conditions, deployment would be a wise choice. We need to understand what intervention options exist and the implications of deploying them.

The potential for international tension over SRM, even over research, is clear, and many questions remain. How would a country react if it suffered a drought or a flood following the deployment of SRM by another country, or group of countries? Would we see sanctions or even war? Similarly, how would the international community react if a country launched a large and secretive programme of research into SRM?

The need for inclusive and adaptive governance of SRM research

With SRM research in its infancy, but interest in the topic growing, now is the time to establish the norms and governance mechanisms that ensure that where research does proceed, it is safe, ethical, and subject to appropriate public oversight and independent evaluation. A transparent and transnationally agreed system of governance of SRM research (including norms, best practices, regulations, and laws) does not currently exist.[1] As knowledge of the technical and governance issues is currently low, an effective governance framework will be difficult to achieve until we undertake a broad conversation among a diversity of stakeholders.

SRMGI's activities are founded on the idea that early and sustained dialogue among diverse stakeholders around the world, informed by the best available science, will increase the chances of SRM research being handled responsibly. While much of the limited research on solar radiation management has taken place in the developed world – a trend likely to continue for the foreseeable future – the ethical, political, and social implications of SRM research are necessarily global. Discussions about its governance should be as well.

By establishing contacts and building international trust via informed dialogue at the level of science academies, academics, NGOs, and other key stakeholders, SRMGI strives to create conditions conducive to international cooperation, rather than those that spawn mutual suspicion, misinformation, and unilateralism. SRMGI cannot predict where these research governance discussions will lead, and a predetermined outcome is not its goal. Instead, the project seeks to build global capacity to engage in an informed debate about if and how to responsibly conduct geoengineering research.

Building on its diverse global network of partners and experts,[2] SRMGI has collaborated with academic and non-governmental organisations to host more than 15 regional workshops around the world on the topic of SRM research governance. Participants at these workshops heard brief presentations introducing SRM science and governance and were asked to give *their* opinions on SRM research and governance, rather than being presented with a point of view on how these issues should be handled. Lively discussions and group exercises explored a range of different possible governance options, without attempting to reach early or forced consensus.

Preliminary lessons from SRMGI for the future of geoengineering governance

Most participants in SRMGI's regional workshops, while displaying a healthy scepticism of SRM technologies, recognised the value of SRM research. In general, they also expressed a high level of interest in international cooperation on low-risk research activities, and in developing international governance arrangements. As one might expect, there was a range of opinions on what forms of governance should apply to different forms of research, what role the United Nations might play, and the desirability of developing SRM technology.

SRMGI discussions also benefited from broad participation by people from diverse regions, sectors (social and natural sciences, policy and legal experts, government, academia, industry, and civil society) and disciplinary backgrounds. SRMGI's affiliation with respected scientific institutions also fostered trust among diverse participants and encouraged open discussion, without the pressure of achieving consensus.

An important element of SRMGI's work has been the decision to avoid identifying preferred or consensus options among different governance arrangements. Instead, SRMGI aims to "open up" discussions of SRM governance by exploring and recording the different perspectives and options that participants express – from no special governance to complete prohibition of research activities. In SRMGI's regional workshops, participants expressed appreciation for the SRMGI approach and pleasant surprise that "experts" did not try to tell them what to think and do about governance of SRM research. This decision to avoid "picking winners" has been seen among both developed and developing country stakeholders as a key component in establishing trust and encouraging participation in SRMGI activities.

Perhaps the most important lesson from the SRMGI experience is that future efforts to engage diverse communities must build on what has come before, without being prescriptive. Disparate, isolated dialogues on SRM research governance risk wasting limited resources while failing to build the cooperative bridges needed to manage potential international conflicts.

Building cooperative bridges: what is needed?

In order to build the capacity for an informed global dialogue on geoengineering governance, a critical mass of well-informed individuals in communities throughout the world must be developed, and they must talk to each other, as well as to their own networks. A transnational conversation about governance regimes may credibly evolve to ever stronger norms through an expanding spiral of distinct, but linked, processes, with agreed arrangements for interaction. For instance, input from SRMGI's global stakeholder discussions could assist with the development of well-connected local expert networks in regions around the world. Indeed, participants in SRMGI workshops in Africa are exploring the possibility of establishing expert working groups and SRM research governance "centres of excellence" in African universities. Efforts to build local capacity will help developing countries to make their own decisions about SRM research governance, informed by their own experts.

In turn, linkages among well-informed local efforts focused on establishing appropriate research governance mechanisms may assist with the cooperative development of a set of "model governance guidelines". These guidelines might identify both substantive and procedural norms for research governance, drawn from transnational stakeholder conversations as well as lessons learned from current geoengineering research activities and controversies. Experience and lessons learned from local efforts to establish research governance mechanisms could also feed back into further development of the guidelines, and assist other regions confronting similar challenges. This "virtuous circle" of learning and collaboration among global stakeholder communities holds the potential to foster the trust needed to manage potential international conflicts over SRM research governance.

SRMGI and its partners continue to bring together researchers, policymakers, and civil society groups to build collaborative bridges across disciplines and regions and to define and establish areas of greatest agreement – as well as persistent differences – in the development of geoengineering research governance. Through regional workshops and broader, cross-regional discussion forums, SRMGI will continue to focus on expanding the conversation about geoengineering research governance where public engagement with geoengineering science and policy is currently low, but potential impacts from geoengineering activities might be high. This will help create the conditions necessary for a considered and cooperative future international dialogue on more formalised systems of governance.

First published online as an opinion article in 2013.

Notes

1 The UN Convention on Biological Diversity (CBD) is the only international instrument that has specifically addressed all forms of geoengineering, but its guidance is general, is not considered legally binding, and only has relevance where proposed activities present a significant threat to biodiversity.
2 Drawn from a variety of disciplines and from 16 different countries, SRMGI's working group helped guide and shape the initiative in its early stages. SRMGI also established a network of partner non-governmental organisations, headquartered in various countries and representing a wide range of viewpoints on geoengineering. Further information on SRMGI's programme of action and meetings can be found on its website (www.srmgi.org).

33

FORESIGHT IN CLIMATE ENGINEERING

Johannes Gabriel and Sean Low

The future paradox in climate engineering

From a futurist's perspective, there are two important dimensions to the discourse on climate engineering. First, climate engineering is a novel enterprise embedded in the complex and dynamic landscape of natural and human systems. This encompasses the entirety of the biosphere and climate system, and an evolving spectrum of states, organisations, and communities, with all of their constituent values, agendas, actions, and effects. Our understanding of this interconnected terrain is inevitably incomplete. This makes it difficult to dissect the climate engineering discourse into distinct elements that can be evaluated independently – and next to impossible to identify and predictably extrapolate trends into the future.

Second, insight into the risks and uncertainties most relevant to decision-making in climate engineering could be gained from observing events that have not happened yet. Models can help us to understand complex environment systems through simulation (Robock and Kravitz, 2018). Small-scale experiments outside the laboratory could help to test certain technical assumptions (Dykema et al., 2014; Keith et al., 2014). Even so, these methods give us no opportunity to observe implications for societal impacts, though we might attempt to circumvent this by examining the history of climate engineering and past emerging technology debates for indications. Ultimately, observing or measuring the impacts upon intertwined physical and social systems would require a global and long-term deployment, which might produce unintended and irreversible outcomes. In other words, in the attempt to learn about potential unintended consequences we might want to avoid, those very outcomes are created.

These premises create a paradox in which we know that we cannot know the future of climate engineering. This is shaky ground for scientists who want to understand the impacts and processes of future climate engineering, stakeholders who want to know the impact it may have on their lives, policy analysts who want to assess and present policy alternatives in which they have confidence, and decision-makers representing broader society who want to set our societies on the path towards a sustainable and secure future.

The basics of foresight

Foresight methods (Glenn and Gordon, 2009) can help to navigate this paradox, for three reasons:

1. Foresight is, first of all, not about predicting the future, but rather about building virtual futures to enrich deliberations and influence policy in the present day.

In this perspective, we will mostly refer to scenario-based foresight methods, which gather deliberately diverse groups of stakeholders and require them to debate and crystallise their combined knowledge into richly descriptive thought experiments and storylines of the future. Of course, such products are not guaranteed realities. The point is to investigate "What if?" rather than predict "What will be?" Ideally, generating such scenarios forces stakeholders to consider alternative developments, linkages, threats, and opportunities – especially ones on the outer edges of plausibility. Foresight thus aims to support decision-making under deep uncertainty by making current plans aware of contingencies that cannot be generated by probability-based methods, and to explore and incorporate points of agreement and contention between stakeholders today.

2. Second, foresight focuses on plausibility as a criterion for thought experiments.

Future developments could become reality as long as one can find rational and internally consistent arguments to explain why something is not impossible. Conversely, game theory and economic modelling, in gauging future strategic actions in climate engineering, assume that actors (e.g. states) act upon probabilistic calculations of self-interest. These frameworks are based on the assumptions that accurate probability functions can be calculated, and that actors will always make rational choices based on such probability calculations. However, the structure of complex, non-linear systems – particularly including human social systems – make such assumptions highly questionable. Plausibility, on the other hand, enables a much wider scope of outcomes, as it only insists that thought experiments have to be logically coherent.

3. Finally, scenarios are generated from mutual learning among diverse stakeholders.

As such, it provides a platform for future-oriented communication. Exploring complex systems and alternative futures requires the systematic integration of scientific, policy, and intercultural perspectives, even emotions. Critical rationalism alone is not enough for a scientific inquiry into the future; it needs self-reflection and even scepticism. Individuals, and even groups unique to a single community or discipline can exhibit cognitive biases that lead to one-sidedness and linear thinking.

Applying foresight to climate engineering

We discuss here a pair of scenario-based foresight projects in climate engineering, in order to demonstrate some aspects of foresight in action.

The *Scenario Planning for Solar Radiation Management* workshop at Yale (September 9–10, 2011) gathered 17 scholars and practitioners from atmospheric science, science and technology studies, and policy, history, economics, ethics, security, and humanitarian aid to explore the question: "What key uncertainties need to be reduced before SRM research and deployment can be considered?" Participants generated and defined a list of key uncertainties, before narrowing them to the inclusiveness or fragmentation of the geopolitical landscape, and the degree of technical knowledge and influence over the direct physical impacts of SRM deployment. These two parameters of uncertainty were further shaped into varying, stylised subsets; these were then developed into scenarios by interdisciplinary working groups (as journalistic articles reporting "from the future"). Four scenarios highlighted imaginative contingencies that might develop alongside decisions to deploy SRM, while two more examined scenarios were built around foregoing SRM, with outcomes depending on the surrounding political and technical landscape (Banerjee et al., 2013).

Another is the *Global Governance Futures (GGF) Program*, which ran over the course of 2014 and 2015.[1] GGF brought together 25 young professionals from China, Germany, India, Japan, and the US to look ahead to the year 2025 and recommend ways to address global challenges in the fields of internet governance, global arms control, and climate engineering governance. The group first engaged in a "horizon-scanning exercise" to collect and define factors that influence climate engineering governance, before ranking their potential in terms of *uncertainty* and *impact* to influence future developments. The group then discussed how factors influence each other via a cross-impact assessment. This was the project's core methodological step, designed to integrate interdisciplinary knowledge and international perspectives, and construct comprehensive and detailed thought experiments envisioning future systems of governance. The scenarios were published in a report, together with policy recommendations on how to handle uncertain threats and opportunities in climate engineering governance (GGF, 2015).

Both the Yale workshop and the GGF programme investigated key uncertainties through mutual and interdisciplinary learning and used them to construct virtual futures that aim to provide insights into the dynamics that might shape the climate engineering debate in particular, and global governance more generally. Some details, however, may have been influenced by the time in which they took place, reflecting evolving landscapes and priorities. The Yale workshop sought to investigate dynamics over a long term of several decades, choosing to sacrifice near-term realism in favour of provoking thought over wide-ranging evolutions in politics, society, and climate. GGF's efforts arrived in a later context where, for instance, governance over field tests has become more prevalent in discussions, and therefore had a more specific focus on the uncertainties influencing nearer-term governance in 2025.

In terms of designing policy-oriented outputs, the earlier Yale workshop was designed to be a standalone exercise that hoped to serve as a pilot project for the use of foresight amongst academics and policymakers in climate engineering. Building from this, the GGF programme linked its scenarios both to an analysis of consequences, and to robust strategic options to showcase how global governance could be strengthened under conditions of uncertainty, but without narrowing the room to manoeuvre in the future. Both efforts provide foundations for further foresight exercises focused on climate engineering.

Foresight for strategic management or community learning?

Given these examples, what might some future uses of foresight in the climate engineering discourse be? At least two objectives are worth brief consideration here.

The first is as a *strategic management* tool: the generation of "aspirational" futures towards which policy can be guided. This is often applied in organisational contexts, for example, amongst institutions and corporations. However, we would suggest that this is not yet a valuable application of foresight in the climate engineering debate. The climate engineering "community" – if one can even be understood to exist – consists of multiple, overlapping research, policy, and civic groupings. Many basic questions have not produced answers with widespread agreement. What are the risks and potential benefits of the various subsets of SRM and CDR? What should be their development pathways? Where and how should they be governed? What are or should be the burdens upon research or technology developers, or the acceptable impacts upon the global environment and the lives and livelihoods dependent on it? A coherent, desired future for climate engineering research would at this early stage be misguided, considering the multiplicity of views.

Rather, we would argue that foresight should for now harness and map this diversity, acting as a tool for *community learning* on such questions, as demonstrated by both the Yale and GGF initiatives. We believe foresight methods should be understood as a set of tools for enabling structured and ongoing social communication, incorporating new voices, and revealing new research and policy gaps. Existing scientific research programmes into the physical and societal implications of climate engineering provide possible sites for deploying foresight tools in this way. This does not preclude policy-relevant research, as long as the intent is to strengthen the resilience of existing efforts in research and governance to future pitfalls and not on setting the climate engineering discourse onto a particular pathway. For an emerging discourse with many unknowns, this is an approach that could help society explore the many different pathways and outcomes, without becoming locked into any one narrow perspective.

Note

1 Visit www.ggfutures.net

34

TOWARDS THE ANTICIPATORY GOVERNANCE OF GEOENGINEERING

Rider W. Foley, David H. Guston, and Daniel Sarewitz

Introduction to anticipatory governance

With geoengineering, we must contemplate a future that is imminent but still uncertain, that is capacious but might not hold what we desire. Above all, we must recognise that "anticipation" is fundamentally ambiguous because what we get may not quite be what we were hoping for. We don't want to rush headlong into a future dominated by either unchecked global warming or the risks of planetary-scale climate interventions rendered real. Yet our ignorance is vast, and our ability to predict is overwrought. So we are left to anticipate.

Anticipation, however, is not purely passive.

This chapter aims to describe what we call "anticipatory governance", which refers most directly to building the capacity to manage emerging technologies while such management is still possible. We test this concept on geoengineering, a set of emerging technologies and techniques. Yet, a similar approach could be used for synthetic biology, biotechnologies, information and communication technologies, or robotics. The category is not limited to contemporary technologies; the railroad in the 19th century, for example, or nuclear weapons in the 20th, might be two historical examples of emerging technologies, even if it is too late to anticipate them now.

The issue with emerging technologies is not that they are new if by new we mean unprecedented. In the field of nanotechnology, chemists, materials scientists, and others have performed research at the nano-scale for decades, while biologists who have performed research under the label of synthetic biology may argue that it is indistinguishable from molecular biology or genetic engineering research. Geoengineering in this sense has a much longer pedigree through the vast changes made by humans to landscapes, coasts, ecosystems, and watersheds, not to mention more recent, explicit efforts to modify the weather (Fleming, 2010).

But having precedent is not the same as being same-old-same-old. Likewise, raising issues of novelty does not necessarily mean raising novel issues – a technology does not have to raise an issue never before seen for it to be novel. Being novel in a particular context is sufficient for serious attention, for example, as even a familiar species in a new environment can become an invasive pest. It is precisely that emerging technologies both have antecedents and are new – that they have what we would call a *politics of novelty* – that makes them interesting and problematic (Rayner, 2004; Guston, 2014).

Using nanotechnology as an example, the politics of novelty revolve around funding. For example, the US National Nanotechnology Initiative (NNI) coalesced research efforts previously conducted in disparate disciplines and enacted favourable budgetary treatment. Nano's politics of novelty includes issues of intellectual property and regulation: if materials and their properties are so novel as to be protected by patents, can they at the same time be similar enough to pre-existing materials at the bulk scale to warrant no special regulatory attention? And issues of risk: if these properties are so novel, why assume that they are benign in the human body or external environment? For synthetic biology, the politics of novelty revolves around the questions of when, during millennia of animal husbandry and breeding plants, do the ambitions of researchers to reconceive of life in engineering terms demand a reconceiving of regulation, governance, and even ethics as well?

For geoengineering, the politics of novelty revolves around the ostensible transition from self-conscious but small-scale interventions in the Earth system, e.g., weather modification of various sorts and environmental restoration and larger-scale but less self-conscious interventions such as industrial-scale agriculture, fossil fuel emissions, and the like, to interventions that would be both large-scale and self-conscious. After we recognise what we have already done (without quite being aware of it) there is no Nature left out there – in Stewart Brand's language, "it's all gardening" but at a planetary scale (Brand, 2009). As this volume so aptly describes, there are other political issues – e.g., the vexing flip side to the collective action problem of climate change in which a single nation or lone private actor could, perhaps, "solve" the problem.

In addition to the politics of novelty, emerging technologies like geoengineering are distinguished by the combination of high stakes and high uncertainty that characterises what Funtowicz and Ravetz (1993) identify as "post-normal science" (Guston, 2014). For geoengineering a major challenge is the radical uncertainty associated with such a technological solution coupled with the high stakes involved (for those who may see themselves as potential losers or winners), and the growing sense of urgency among some scientists and activists. Post-normal science explicates the challenges for "building knowledge" that:

i.   Appropriately accounts for the societal and technical dimensions of societal problems
ii. Directly contributes to problem solving

iii.   Legitimately includes diverse types of knowledge, and

iv.   Credibly connects to the local contexts that inform socio-technical problems" (Bernstein et al., 2014, p. 2493).

Those aspects invite a new set of operating conditions for geoengineering research, which includes "extended peer review" or public engagement in decisions otherwise left to scientists alone in normal science (Funtowicz and Ravetz, 1993). Such engagement may be relatively spontaneous, arising from the publics themselves, as with public scepticism and opposition to nuclear power in the 1960s to embryonic stem cells in the 2000s.

Alternatively, public engagement may be premeditated and even somewhat contrived by government or civil society groups, for example via consensus conferences aimed at eliciting public views through formal deliberative activities. Anticipatory governance builds upon such activities by envisioning two additional capacities: anticipatory knowledge that does not seek to make predictions or probabilistic forecasts; and integration of social scientific and humanist knowledge with practices with natural science and engineering research (Barben et al., 2008; Guston, 2008; Sarewitz, 2011; Guston, 2014).

This chapter next explicates a rationale for anticipatory governance and its three conditions: foresight, engagement, and integration. Next we test its use as an evaluative tool for discursive and governance processes by investigating five early reports on geoengineering – including geoengineering's own Asilomar meeting. Our analysis offers evidence that elements of anticipatory governance are in practice and that we may, in fact, be doing better than we have done recently with nanotechnology or with genomics; however, work remains to be done to clarify and specify the conditions for practising anticipatory governance well enough to realise the ideals of deliberative democracy to reflect the values and capabilities of pluralistic societies. Yet, "well enough" would not mean the "control" of emerging technologies, an impossible and even incoherent goal (Stirling, 2014).

Why anticipatory governance?

Anticipatory governance is a vision for dealing with emerging technologies by building the capacity to manage them, while management remains possible. It stands upon two concepts: One, governance, is a broad-based societal capacity to make collective decisions. It is not synonymous with, but includes, government action (e.g., treaties or legislation) as well as non-governmental action (e.g., market pricing or protests) and activities that require public-private collaborations (e.g., standards) or that occur in both public and private contexts (e.g., funding and subsidies, insurance and indemnification). Guston points out that taking a governance perspective does not promote a neo-liberal agenda, but recognises that the "complicated political economy of technoscience" cannot be squeezed into crude dichotomies like government versus market or promotion versus banning (Guston, 2014, pp. 226–227).

Two, anticipation, expresses a particular kind of disposition towards the future governance. It is our perspective that all governance requires some explicit disposition towards the future, but anticipation stands apart from, say, prediction or precaution. An anticipatory disposition is not about seeing into the future (prudence) or saying what the future is going to be (prediction) or estimating the chances of a certain outcome (probabilistic forecasting), all of which prescribe a "knowledge first" approach to action (Sarewitz, 2011). Rather, from Latin for "prior" and "capacity", anticipation is about doing something now, like building a capacity, in preparation for something that might occur in the future. Similar to precaution, it recognises a radical uncertainty. But it differs from precaution because precaution implies some difficult-to-specify demarcation between a state of action and a state of inaction, and often acknowledges an easing of uncertainty based on more research or more data, which anticipation does not (Dupuy, 2007). Precaution prescribes a waiting game that is potentially self-defeating, as new information may never resolve uncertainty, and waiting may forego benefits derived from innovation.

The Center for Nanotechnology in Society at Arizona State University (CNS-ASU) adopted anticipatory governance as a strategic vision due to shortcomings with predictive and precautionary approaches. This led CNS-ASU to design research programmes that helped develop the capacities of foresight, engagement, and integration. While CNS-ASU has popularised anticipatory governance for nanotechnology and others have begun to discuss it for synthetic biology (Gorman, 2012) and geoengineering (Macnaghten and Szerszynski, 2013), still others have discussed it more broadly. As anticipatory governance spreads through academic literature and into practice, it is worthwhile reflecting on its origins, its reasons for being, and in response to those like Global Economic Forum,[1] its authentic tenets.

Karinen and Guston (2010) and Guston (2014) explain that while the origin of the term itself is largely obscure, scholars began using anticipatory governance in a coherent way about 2000, in public administration, environmental policy, and soon thereafter in nanotechnology. Yet one may trace the conceptual roots of anticipatory governance back to Alvin Toffler's best-selling *Future Shock* (1971), which introduces the cognate term of "anticipatory democracy". He articulates a marriage of New England-style participatory democracy to the somewhat novel (however, technocratic) means of strategic and long-range planning, budget forecasting, and the like. Toffler's rationale for anticipatory democracy is cognate as well; he intends it to be therapeutic for "future shock", a societal malady induced by rapid technological change, about which Toffler harboured some seemingly determinist views.

Toffler does not specifically contemplate integration of social science into natural science to slow or steer innovations responsible for future shock. Perhaps ironically, integration can trace its genealogy back to post-World War II policy, as the United States was refashioning its research enterprise from its wartime exigencies. Politicians and scientists battled over such issues as democratic accountability of civilian research, intellectual property, and (pertinent to this discussion) the role of the social sciences. Many elite natural scientists believed the social sciences were technically and perhaps ideologically suspect, and thus should be left out of publicly funded research for fear of tainting the whole endeavour. Detlev Bronk – who would go on to become president of the National Academy of Sciences, Johns Hopkins University, and Rockefeller University – voiced support for the social sciences. While his sentiments were instrumentally oriented, Bronk nevertheless argued that "[c]ompetent social scientists should work hand-in-hand with natural scientists, so that problems may be solved as they arise, and so that many of them may not arise in the first instance" (Bronk, 1975, p. 413).

Bronk did not win the day, and the social sciences needed another generation – till Toffler's time – to receive but junior membership in the formal scientific establishment. At approximately the same time, Congress created the Office of Technology Assessment (OTA) in 1972. However, the OTA tilted in a distinctly technocratic direction and away from Toffler's participatory impulse (Sclove, 1995; Bereano, 1997). The OTA also never quite managed to develop aspects of technology assessment that dealt with foresight, due largely to the purview of its congressional client. Its work essentially became policy analysis with a particular focus on science and technology (Bimber, 1996). Not until two decades later when the US Human Genome Initiative created its programme on the ethical, legal, and social implications (ELSI) of genomic research, did a research programme acknowledge that funding societal research might be a wise complement to potentially transformative science. The separation between "real science" and ELSI was an explicit part of the deal, and there was little thought that insights emerging from ELSI would (or should) feed back into policy processes steering genome science (Cook-Deegan, 1994).

After the closure of OTA and the modest achievements of the genome ELSI programme (McCain, 2002), Guston and Sarewitz (2002) called for a more decentralised, more constructive (e.g., Schot and Rip 1997), more participatory technology assessment that could support anticipatory governance. While science and technology studies was late to the party (Bennett and Sarewitz, 2006), social science publications addressing nanotechnology have increased rapidly since 2006 (Shapira et al., 2010). This increase is fuelled by mandates in many countries to include social and ethical research along with nano-scale science and engineering research. In some cases, societal research on nanotechnology, as in CNS-ASU or scholars in the United Kingdom working on responsible innovation (e.g. Stilgoe et al., 2013a), built capacity to address geoengineering.

We cautiously suggest, then, that starting in the mid-1970s there has been some evolution of theory, practice, and policy towards an explicit commitment to anticipatory governance to address the uncertain futures of emerging technologies. The main evolutionary threads are: 1) a continual distancing from the naïve belief that the future of technologies-in-society is predictable and can be governed as such;

2) an increasing commitment to more formal mechanisms of public participation in both anticipation and governance of emerging technologies; and 3) an increasing role for social science through integration with natural science in seeking to come to grips with the socio-technical complexities of emerging technologies.

Anticipatory governance in action?

What looks to be an emergent, if informal and unorganised capacity for anticipatory governance appears to be coalescing around geoengineering. Scientists and science organisations, non-governmental organisations, and government bodies are applying their capacities to begin imagining what geoengineering might be like, planning what research might be necessary to achieve (or avoid) imagined futures, and what institutional designs and guidelines might be necessary to govern research and possible deployment. These early efforts include the vocal presence of social scientists and humanists and disparate activities aimed at understanding public attitudes about geoengineering and engaging publics in discussions about it. All this has occurred despite the paucity of technical capacity to consciously geoengineer the climate.

Reasons for this activity seem apparent: the risks and uncertainties of trying to manipulate the global climate through conscious technological intervention are enormous. Scientists are quite reasonably uninterested in shouldering the burden of these risks and uncertainties alone. Many scientists have proven more than willing to share this burden with policymakers and even the public.

Yet a strong, countervailing force has been afoot. This force is conspicuously present in the allocation of huge intellectual, technological, and financial resources to advance scientific knowledge about climate change, and especially to predicting the future evolution of the climate. Here scientists promised that research would reduce uncertainties and predict the future of human activity on the behaviour of the atmosphere as a foundation for subsequent policy decisions. The idea was obvious and largely uncontested: fundamental scientific understanding aimed at a predictive understanding of anthropogenic climate change will motivate governments to intervene through public policies that will reduce the consequences of climate change, mostly by reducing greenhouse gases emissions (Pielke, 1995). And while the uncertainties of climate change science are now a fervid political battleground, a larger point is generally neglected: regardless of how certain scientists are about climate change, the future risks and uncertainties of trying to manipulate the global climate through policy action are no less enormous or daunting than they are with geoengineering – they are just different.

The larger contrast here that we want to emphasise is between prediction that depends on conventional (if discredited) notions of "rational" decision-making – predict the future of human impacts on climate, then act on the basis of those predictions (e.g. Lempert et al., 2004) – and an anticipatory approach that sees the future not as something that can be predicted but as something to be made through encounters among pluralistic worldviews, political action, technological change, and so on. What anticipatory governance does is try to condense and make explicit some of those encounters, in the context of decisions around the evolution of a particular technology or technological system.

We offer this contrast to make a key point about both geoengineering and anticipatory governance. Interminable and largely counterproductive political debates around climate change policy have been stunningly detached from consideration of the actual impacts of such policies on the future of climate (e.g. Sarewitz and Pielke, 2008). Perhaps this detachment is because the causal chains from economic and policy incentives to reduced use of fossil fuels to mitigate climate risks are too complex, attenuated, and uncertain to provide good boundaries for arguments about climate policy. As Pielke et al. (2000) discussed, there is no "big knob" for dialling down climate impacts to some specified level through particular policy interventions. Geoengineering, in stark contrast, grabs attention precisely because the very idea (whether plausible or mere folly) posits a direct link between a fairly constrained set of actions and a climatic consequence.

To be clear, we are not offering a brief on behalf of geoengineering, but rather on behalf of the possible salutary effects of the anticipatory governance of geoengineering on the climate debate more broadly. Anticipatory governance provides a set of tools for focusing that attention in ways that are unavailable in the larger climate change discourses. The idea of geoengineering technologies disciplines the public and the scientific imagination in a way that predictive science cannot.

Anticipatory governance as an evaluative tool

In addition to providing strategic vision, anticipatory governance can be an evaluative tool, using criteria derived from its normative perspective (Barben et al., 2008). Here we evaluate the processes and recommendations of five early efforts to explore geoengineering governance that issued written reports published in 2009–2010. These reports occurred in a moment of time when the governance issues of geoengineering were taken up and deliberated by diverse stakeholders. This time frame offered a window of opportunity for the concepts of anticipatory governance to effect changes in issue framing and decision-making. We ask the following questions drawn from the anticipatory governance approach about the processes that created the reports:

a. Foresight – What influenced considerations of the future? What methods were used and how did this framing influence the process?

b. Engagement – Who was engaged? Whose interests were represented? How transparent were the proceedings?

c. Integration – Were diverse knowledge types, e.g. social and natural sciences or policy and engineering, integrated in the process?

d. Ensemblisation – Were steps taken to harmonise these three elements throughout the process?

We then evaluate the reports' recommendations:

- To what extent does the report adopt an anticipatory (rather than predictive) disposition towards the future?
- To what extent does the governance proposal involve substantive engagement with publics?
- Is engagement considered to be traditional science communication to fill the knowledge deficit of the public or to shape research agendas and inform scientific inquiry?
- To what extent does it attempt the integration of knowledge across disciplinary (and other) divides?
- Is the approach to integration hierarchical, separate but equal, or mutually collaborative?
- To what extent are these capacities brought together (e.g., such that anticipatory activities are also well integrated and participatory)?

Our analysis looks primarily for high-level policy goals and secondarily for demonstrations of "on-the-ground" mechanisms to operationalise these policies. Both are required, of course, because high-level policy goals can be on target, for example, while the specific instructions for implementation are inadequate or contradict the goals. Specific articulations suggest how the policy goals will be interpreted and, thus, implemented.

Efforts towards governance of geoengineering

The Royal Society produced *Geoengineering the Climate: Science, Governance and Uncertainty* in September 2009. It is chronologically the first report, and its influence is evidenced by the direct citations in the other four reports (see Table 34.1). The Royal Society's report defined geoengineering as "deliberate large-scale intervention in the Earth's climate system, in order to moderate global warming" (Royal Society, 2009, p. ix). The report sought to clarify "scientific and technical aspects of geoengineering" with an aim to contribute "to debates on climate policy" (p. v). The Royal Society defined two mechanisms for geoengineering: carbon dioxide removal (CDR) and solar radiation management (SRM).

The UK House of Commons Science and Technology Committee (HOC S&T) issued *The Regulation of Geoengineering: Fifth Report of Session 2009–10* with the explicit goal of addressing three regulatory-focused questions for the agencies administered by the prime minister and government ministers. The HOC S&T adopted the terminology and definitions offered by the Royal Society. The HOC S&T and the US Government Accountability Office (USGAO) formed a bilateral international collaboration to generate complementary reports on geoengineering governance. As shown in Table 34.1, there is evidence of cross-referencing between the HOC S&T and USGAO reports.

TABLE 34.1 The cross-influence between selected geoengineering reports

Process Evaluation	Royal Society	HOC-STC	USGAO	BPC	CIAG	Influenced (total)
Royal Society	X	0	0	0	0	0
HOC-STC	93	X	13	0	1	107
USGAO	33	2	X	0	2	37
BPC	3	0	1	X	0	4
CIAG	9	1	7	0	X	17
Influential (total)	136	3	21	0	3	

Note: Reports listed in the left column are in chronological order. Direct references or testimonies from other reports are tabulated along each row. The bottom row shows the total instances that a report was referenced, a measure of its influence on the other reports. The right column sums instances in which a report relied on another, a measure of how it was influenced.

The USGAO report, *Climate Change: A Coordinated Strategy Could Focus Federal Geoengineering Research and Inform Governance Efforts*, adopted the Royal Society's definition, but highlighted the uncertainty and complexity in defining geoengineering, for "without the guidance of an operational definition … agencies may not recognize or be able to report the full extent of potentially relevant research activities" (USGAO, 2010b, p. 23). The USGAO report, requested by the chairman of the Committee on Science, Technology in the House of Representatives, investigated the state of the science, the current efforts of the US in geoengineering, and views of experts on the regulation and governance.

The US-based Bipartisan Policy Center (BPC) issued *Geoengineering: A National Strategic Plan for Research on the Potential Effectiveness, Feasibility, and Consequences of Climate Remediation Technologies* to build upon work by the National Academies of Sciences (2009). The BPC report explicitly avoided the term geoengineering, preferring "climate remediation" to describe "technologies that are intentionally designed to counteract the climate effects of past greenhouse gas emissions to the atmosphere" (BPC, 2012, p. 3). The BPC report references CDR and SRM to enact climate remediation. It offers an initial strategy for the US government to "go about improving its understanding of climate remediation options and how it should work with other countries to foster procedures for research [and] is offered as an exploration of what might be appropriate responses to changes in the global climate measured in recent decades" (BPC, 2010, pp. 3–4).

The Climate Institute Asilomar Group (CIAG) issued *The Asilomar Conference Recommendations on Principles for Research into Climate Engineering Techniques* for the scientific community as well as for governments and civil society worldwide. The goal was to "initiate a broad, interdisciplinary dialogue among experts that would produce guidance for the scientific community to responsibly and safely develop, test, and evaluate the potential for intentional intervention in the climate system [and] provide input for consideration of necessary and optimal mechanisms for planning, conducting, and overseeing scientific research"

(Asilomar, 2010, p. 14). The CIAG used the term *climate engineering* to "refer to activities taken to counterbalance global warming and its impacts" (p. 12) and described *remediation technologies* in reference to CDR and *intervention technologies* for atmospheric alterations, such as SRM.

This section presents our process evaluation of the five reports. Overall, the processes that led to the reports range from the small, exclusive BPC group to the more than 165 participants at the Asilomar conference. Each considered the future implications of climate change as the driving force behind their work, and several addressed future societal dynamics. The BPC and HOC S&T processes integrated diverse knowledge types, suggesting awareness that each can benefit from the other, what we term here "mutually supportive", while the Royal Society and CIAG attribute equal value to different knowledge types but do not seek to integrate them. The USGAO process was so opaque that it was not possible to discern how diverse knowledge was integrated. Each struggled to blend ("ensemblise") foresight, engagement, and integration.

Royal Society: Geoengineering the Climate

John Shepherd, fellow of the Royal Society and professorial research fellow in Earth system science at the University of Southhampton, led the 11-member working group. A four-member science policy team and a seven-member review panel supported the working group. The report's predisposition to the future technology made explicit reference to the "technology control dilemma" (Collingridge, 1980 as cited in Royal Society, 2009, p. 37). The issuing of public statements from 51 individuals and 26 organisations illustrates the representation of contributors (see Table 34.2). The responses to the public call are available at an online repository.[2] The report integrated a diversity of disciplinary knowledge that we characterise as separate but equal. The report's structure reinforces separation between technical and social issues. As such, the sections sometimes take contradictory stances, exemplified by a focus on modelling in the introduction that stands in contrast to the "control dilemma" expressed in section 4.2. While the three elements of anticipatory governance are present, they are uncoordinated.

HOC-STC: The Regulation of Geoengineering

Member of Parliament Phil Willis chaired the HOC S&T (see Table 34.2). Geoengineering was framed as "Plan B" to climate change mitigation and adaptation. Societal dynamics are implicit, at best, in the discussion of regulatory responses. Expert testimony at both ends of the pro-con spectrum contributed evidence. Moderate social scientists and climatologists offered oral testimony and integrated comments on all topics of the inquiry. However, the report separates out these topics. The report's recommendations attempt to harmonise elements of anticipatory governance.

TABLE 34.2 Report process analysed with anticipatory governance

		Royal Society	HOC-STC	USGAO	Asilomar	BPC
Engagement	Author	J. Shepherd & 11 members	P. Willis & 13 members	F. Rusco & 16 staff	M. MacCracken & 13 members	J. Long & S. Rademaker & 16 members
	Gender ratio (M:F)	9:3	12:2	9:8	13:1	17:1
	Transparency	All material public	All material public	Testimony private, no attribution in text	Testimony private, attribution in text	Closed meetings, report open access.
	Public	Yes = n14	Yes = n2	No	No	No
	NGO	Yes = n11	Yes = n1	Yes = n2	Yes = n25	Yes = n3
	Media	No	No	No	Yes = n18	No
	Corporate	Yes = n5	No	No	Yes = n13	Yes = n1
	Academic	Yes = n35	Yes = n8	Yes = n9	Yes = n84	Yes = n9
	Government	Yes = n12	Yes = n2	Yes = n13 agencies	Yes = n31	Yes = n5
	Legal Advisors (non-academic)	Yes = n4	No	No	Yes = n4	No
	Private investor	No	No	No	Yes = n8	No
Foresight	Predisposed to future, if so how?	Yes, "Anticipating in the early stages how a technology will evolve is difficult" (p. 37).	Yes, "as part of a portfolio of responses to climate change" (p. 6).	Yes, lack of scientific knowledge creates uncertainty. This needs to be resolved.	Yes, positioned to build upon previous reports.	Yes, "parameters must change over time as understanding of the risks of climate remediation evolves" (p. 3).
	Societal dynamics	Yes, addresses the "reversibility of society's commitment to a technology" (p.37).	Yes, but implicitly within the framing of regulatory needs.	Yes, addressed in section on political, economic, and ethical concerns.	No, "understand potential responses to [...] more completely" (p. 4).	

(continued)

TABLE 34.2 Continued

		Royal Society	HOC-STC	USGAO	Asilomar	BPC
Integration	Integration	Yes, "geoengineering will be determined as much by social, legal and political issues as by scientific and technical factors."	Yes, contributions are integrated with social scientists and climatologists commenting on all topics of the inquiry.	No, expert testimony integrated by USGAO staff, not entirely transparent process.	Yes, "Their expertise covered Earth, environmental, and social sciences, risk assessment, public policy, ethics, philosophy, history, economics" (p. 7).	Yes, "scientific, science policy, foreign policy, national security, legal, and environmental communities … a wide range of perspectives and expertise to the task force" (p. 2).
	Diverse form	**Separate but equal**	**Mutually supportive**	**Unknown**	**Separate but equal**	**Mutually supportive**
Ensemblisation		No, segmentation of key concepts.	Recommendations present a list of mutually supportive activities.	No, segmentation of key concepts.	No, division of labour among expert groups.	Collaborative authorship, but not uncontested and harmonious.

USGAO: Climate Change: Geoengineering Research

Frank Rusco, director of USGAO's Natural Resources and Environment division and staffers compiled the report with input from agency and domain experts (see Table 34.2). The lack of scientific certainty as something to be solved was the focus, thus adopting a "normal science" perspective, in contrast to our view of geoengineering as inescapably "post-normal". Minor attention was paid to societal dynamics in the section on political, economic, and ethical concerns. The report uses "experts" to mask attributions from specific persons. There was no public solicitation for contributions. A diversity of disciplines offered testimony, but testimony was segregated to specific topics. There is little evidence that the three components of anticipatory governance are considered together.

BPC: Climate remediation

Jane Long, associate director-at-large of Lawrence Livermore National Laboratory and Stephen Rademaker, former assistant secretary of state, co-chaired the task force. The task force took a predisposition to future technological, scientific, and societal dynamics. Meetings were only open to invited parties, yet the report is publicly available and a press conference was held. The task force integrated knowledge from a wide diversity of perspectives. Contributions were collaborative in the introduction and conclusion with specific topics reported in the report's body. Authorship was collaborative, and the recommendations indicate group consensus, but the group's deliberations were often contentious, illustrating challenges to bringing elements of anticipatory governance together harmoniously.

CIAG: Climate engineering

Michael MacCracken, chief scientist for climate change programmes at the Climate Institute with 25 years' building climate change models led 13 members of the Asilomar Scientific Organizing Committee. The meeting's framing created a knowledge-first approach to climate engineering, expressed throughout the report. Conference organisers invited a diversity of experts who revised a conference statement in an attempt to integrate the diverse perspectives, yet the agenda reflects a segregation of knowledge domains.

Evaluating recommendations with anticipatory governance

Evaluating the reports' recommendations against anticipatory governance, we find that all five articulate foresight, engagement, and integration as high-level principles. However, three of the reports express foresight as enhanced prediction, engagement as traditional science communication, and integration as either hierarchical or separate but equal. As such they stand at odds with our view of anticipatory governance. The Royal Society and CIAG elicited broad represen-tation of views and diverse knowledge types during their respective processes, yet neither articulates core concepts from anticipatory governance in the specific recommendations. At the other end of the spectrum, the BPC report is the most exclusive and least open process, and yet its recommendations articulated each and every aspect of anticipatory governance, thus putting processes at odds with the outcomes.

Royal Society: Geoengineering the Climate

The Royal Society report expresses the elements of anticipatory governance at a high level, but calls for prediction (and not exploratory foresight), science com-munication, and dialogic public engagement (an internal inconsistency), and separate but equal (and not collaborative) work between the natural and social sciences (see Table 34.3). The Royal Society (2009) articulates the highest pri-ority for foresight as in predictive models of Earth's climate, "unintended envi-ronmental effects should be carefully assessed using improved climate models as well as the best now available" (p. x) and "detailed modelling of their impacts on all aspects of climate (including precipitation patterns and monsoons) is needed" (p. xi). The Royal Society (2009) certainly addresses engagement as a guid-ing principle for the governance of geoengineering in the form of "stakeholder engagement and a public dialogue process" as a high-level goal (p. xi). The principle is upheld in certain statements such as, "diverse publics and civil society groups could play a much more positive and substantive role in the development of the field, by contributing to analysis of the social, ethical and equity basis of geoengineering proposals" (p. 42). But engagement does not extend beyond the social sciences, as "[p]olicymakers need well-informed and authoritative advice based on sound science", contradicting the goal of public engagement (p. 6) and thus demonstrating internal inconsistencies in the report. There is evidence of integrating diverse knowledge into the report, yet these efforts keep things sepa-rate but equal, at best. The Royal Society advocates for "an international body such as the UN Commission for Sustainable Development" to lead the govern-ance regime. There is little evidence supporting the ensemblisation of foresight, engagement, and integration for governance. The reliance is on clearly dividing labour and minimising feedback or cross-cutting approaches.

HOC S&T: The Regulation of Geoengineering

The HOC S&T's recommendations address the core elements of anticipatory governance at a high level, and its specifics call for exploratory foresight and substantive engagement; nevertheless, knowledge integration remains separate but equal and not collaborative (see Table 34.3). The report explores future regulations and calls for something between moratoria and determinism. For example, "geoengineering is not sufficiently advanced to make the technology

TABLE 34.3 Report recommendations analysed with anticipatory governance

	Engagement	Foresight	Integration	Ensemblisation
Royal Society: *Geoengineering the Climate*	Principle present: Yes Specifics aligned, **No**	Principle present: Yes	Principle present: Yes	Principle present: **No**
HOC-STC: *The Regulation of Geoengineering*	Principle present: Yes	**Predictive** Societal dynamics? **Yes**	**Separate but equal**	Specifics aligned, **No**
	Specifics aligned, **Yes**	Principle present: Yes	Principle present: Yes	Principle present: **No**
USGAO: *Climate Change: Geoengineering Research*	Principle present: Yes	**Exploratory** Societal dynamics? **Yes**	**Separate but equal**	Specifics aligned, **No**
	Specifics aligned, **No**	Principle present: Yes	Principle present: Yes	Principle present: Yes
BPC: *Climate remediation*	Principle present: Yes	**Predictive** Societal dynamics? **No**	**Hierarchy**	Specifics aligned, **No**
	Specifics aligned, **Yes**	Principle present: Yes	Principle present: Yes	Principle present: Yes
Climate Institute: *Climate engineering*	Principle present: Yes	**Exploratory** Societal dynamics? **Yes**	**Mutually supportive**	Specifics aligned, **Yes**
	Specifics aligned, **No**	Principle present: Yes	Principle present: Yes	Principle present: **No**
		Predictive Societal dynamics? **No**	**Separate but equal**	Specifics aligned, **No**

predictable, but this itself is not grounds for refusing to develop regulatory frameworks or for banning it" (HOC, 2010, p. 49). Particular attention is paid to engagement by the report and reiterated in five specific recommendations. The report calls for a proactive approach to governance. However, the integration of diverse knowledge remains separate but equal, as "[d]ecisions [are] to be based on the best scientific evidence, including social science" (p. 52). The report suggests the United Nations Framework Convention on Climate Change should be the organising regulatory body. Elements of anticipatory governance are considered in isolation.

USGAO: Climate Change: Geoengineering Research

The USGAO's recommendations attend to anticipatory governance at a high level, but the specifics call for predictive foresight (and not exploratory), science communication (and not engagement), and separate but equal (and not collaborative; see Table 34.3). The report foresees that geoengineering "may have unintended and significant impacts within and beyond national borders" (USGAO, 2010b, p. 16), but it seeks predictive models to address uncertainties and "inform societal debate and decision-making over what would constitute a 'climate emergency' and whether deployment of a geo-engineering approach would be merited" (p. 16). There are calls for public engagement throughout the USGAO report, but the specifics are off base: "Answers to these [unresolved scientific questions] will also inform the public debate" (p. 38). Diverse knowledge is integrated into the report, but there is a clear hierarchy in the prioritisation of certain knowledge, "Better understanding of the climate and a way to determine when a 'climate emergency' is reached" (p. 16). The USGAO (2010b) adopts the National Research Council's recommendation for "basic climate science research, including (1) improved detection and attribution of climate change to distinguish the effects of intentional intervention in the climate system from other causes of climate change, and (2) information on climate system thresholds, reversibility, and abrupt changes" (p. 16). The USGAO advises that the US Office of Science and Technology Policy (OSTP) coordinate federal research, but the report's representation of engagement, foresight, and integration does not suggest any ensemblisation.

BPC: Climate remediation

BPC's recommendations demonstrate the core elements of anticipatory governance at a high level and, uniquely, in its specific recommendations. BPC calls for exploratory foresight, substantive engagement, and mutually collaborative approaches to integration (see Table 34.3). BPC (2010) recommends that OSTP coordinate federal research efforts, as it is "the only entity in the federal government in a position to realistically coordinate this research enterprise and navigate the technical and political challenges" (p. 17). The BPC report recommends an exploratory approach to foresight. "The environmental, scientific, technological, and social context for climate remediation research is likely to evolve significantly over time in unpredictable ways. Federal research programs should be required to review those changing conditions on a regular basis" (BPC, 2010, p. 14). This quote also highlights the mutually supportive approach to integration. The report addresses engagement early on, as "climate remediation techniques will require new governance structures to engage the public and to set parameters" (p. 3) and throughout the text, e.g., "Robust and durable mechanisms for public engagement should be established early in the research programs" (p. 14). The BPC report (2010) is peppered with references to knowledge integration and identifies other attempts that fail to do so, e.g., "There is also a clear need for a more extensive integration of social sciences than has been achieved so far under either the USGCRP [US Global Change Research Program] or USCCTP [US Climate Change Technology Program]" (p. 18).

CIAG: Climate engineering

The Climate Institute's report articulates anticipatory governance at a high level, but the approach to foresight is predictive, engagement is secondary to scientific knowledge, and integration is separate but equal (see Table 34.3). Foresight is expressed as a means to predict future outcomes, e.g., "[n]umerical modelling studies of the range of approaches that could contribute to moderating climate change and its impacts" (Asilomar, 2010, p. 21). The report opens with the five Oxford Principles including, "[p]ublic participation and consultation in research planning and oversight, assessments, and development of decision-making mechanisms" (p. 9). The report's tenor suggests that laboratory and modelling activities require "business as usual" approaches to governance and that by keeping dangerous technologies in the laboratory, society is assured safety. This perspective influences the approach to engagement, such that the public need not be involved until, "[f]or field experiments, the need for public consultation, like the need for other elements of legitimate governance, will increase with the scale and potential risks of the proposed research experiment" (p. 23). The demand for societal, ethical, or legal aspects is not integrated (and actually excluded) from early-stage scientific research and experimentation in the lab: "[m]odelling and laboratory studies pose little to no risk of impact to the climate, environment, or society, and so new governance mechanisms are not likely to be needed" (p. 18). This perspective reinforces the notion that it is fine for dangerous research to happen in the laboratory, i.e., keep the door closed to protect society, and restricts opportunities to open up the laboratory as a place to explore questions of societal or ethical implications through reflexive inquiry (Fisher, 2014). Yet, the laboratory is often identified as a space for those questions to arise earlier, rather than waiting for the science to emerge from the laboratory and then become subjected to ethical and societal inquiry. With respect to ensemblisation, there is a strong tension between governance and impartiality, "a critical function of national and international governance systems must be to organize and manage competent, impartial, independent, and transparent expert assessments of the benefits and risks of proposed climate-engineering approaches" (Asilomar, 2010, p. 22). The report thus adopts the "normal science" risk-benefit paradigm, which assumes that impartiality is practically and theoretically possible and that values are satisfactorily subject to quantification, even if, as seems likely, uncertainty remains high and values remain strongly contested.

Discussion and conclusion

The core concepts of anticipatory governance are permeating the science-policy dialogue in areas other than nanotechnology. Yet while the high-level goals stated in each report align with anticipatory governance, the specific articulations of the key concepts are rarely aligned. Our evaluation of each report's recommendations highlights areas of divergence from and convergence with anticipatory governance. Geoengineering might be simply a Rorschach test for entrenched positions on climate change. For example, calls for large, international geoengineering efforts by the Royal Society, HOC S&T, and CIAG reports seem to recapitulate the reliance on a dysfunctional global governance regime for climate change. Alternatively, geoengineering is positioned as a "Plan B" made necessary by the apparent intractability of climate change when viewed as a collective action problem. All five reports attend to both CDR and SRM techniques, yet most concerns focus on SRM, as it is "high leverage" – meaning faster and less expensive to deploy. The reports often frame planning for the future as a response to the implications of climate change, rather than about what we want the future to be and what options are open to complex societies navigating an uncertain future. These forums might have been a great opportunity to deliberate on how to understand the relations between human prospects and changing climatic conditions, rather than suggesting that pre-industrial conditions are "natural" and thus desirable.

Some scientists are demonstrating a willingness to share the burden of governance with industry, policymakers, and the public. For example, the Climate Institute's Asilomar report (2010) states, "climate engineering would be much more than a purely scientific decision" (p. 22). However, many scientists argue for the status quo, i.e. no or minimal oversight for laboratory research, even if risky outcomes or technologies are imminent. This notion is exemplified as, "Modelling and laboratory studies pose little to no risk of impact to the climate, environment, or society, and so new governance mechanisms are not likely to be needed" (Asilomar, 2010, p. 18). In such cases, certain scientists want to assume the responsibility for keeping dangerous knowledge and technology in the laboratory, safe from society, rather than opening up to reflexive questions regarding the nature and intent of the research. Yet, if scientists are explicit in wanting to share the risks of geoengineering governance with broader constituencies, a question left implicit is whether there are experiments that are better left undone, and if so, who should get to answer. A fundamental challenge remains, how to bypass the "knowledge first" approach and initiate engagement to proactively explore broader public values.

There is a tension between the process that created the reports and their recommendations. BPC was perhaps the least transparent and inclusive of those investigated, more closed, it seems than the Royal Society. Nevertheless, BPC's recommendations are most directly aligned with anticipatory governance. Any reflective approach would not abandon the integrity of the process so quickly simply upon seeing such results. The obvious explanation is not only that one of this paper's authors (Sarewitz) was on the small BPC task force, but as well that that group included a number of members who were highly attuned to the value of public deliberation and the limits of normal science in addressing complex socio-technical problems *and* that BPC deliberations did not explicitly privilege natural science expertise over social science. At Asilomar, in contrast, the number of participants and privilege to natural scientists may actually have acted against adopting a mutually supportive approach to knowledge integration. Perhaps the broad representation at Asilomar created a paper-thin smattering of alternative perspectives, thus leaving MacCracken and the Asilomar Scientific Organizing Committee the unenviable task of marrying a hodgepodge of perspectives with a less insistent demand for a unified, integral whole than was present at BPC and its focus on consensus.

Each report calls for public involvement, yet the capacity to conduct such involvement in an international setting is meagre. The international governance process employed for climate change, to date, is characterised as dysfunctional (Bodansky, 2013). These five reports highlight complications for engagement between two of the wealthiest and most technologically advanced societies (US and UK). The first recommendation listed in the HOC report is: "We welcome the review that the House is carrying out of the audio-visual facilities in committee rooms to enable the taking of oral evidence in committee by video link". We cannot reasonably expect to engage in a global dialogue if the House of Commons struggled to video-conference with one person (even five years ago). Thus, challenges to global engagement highlight both social challenges with an approach lodged in global governance, as well as for the communications systems for international debate and dialogue that supports it. One possible model for broader participation is the World Wide Views process, organised by the Danish Board on Technology, which conducted public engagements for global warming in 2009 (Chhetri and Grossman, 2012) and biodiversity in 2012 (Chhetri and Farooque, 2012) at numerous sites around the globe. This approach is still rough around the edges, especially in eliciting public values, and has little capacity to translate findings into action or even high-level dialogue.

Several of the reports aspire to create predictive models so accurate that we might predict the next rain cloud to appear over Eeyore's head, and yet there is no clear sense in which such certainty is inevitable, or if inevitable then timely, or if timely then even helpful in resolving governance issues. Only by avoiding the "knowledge first trap" when anticipating risks (Brown, 2009) can we do our best to assure that knowledge creation is responsive to governing needs and that governance issues get the attention they deserve from the start.

The concept of knowledge integration, while apparent in all three reports, remains largely in the realm of "separate but equal" – which is even in this analogy inherently unequal. The division of labour is clear – natural scientists will discover the wonders of nature and social scientists will uncover, later, if people like the discovery. Rather than ascribing to a division of labour there are efforts to integrate diverse knowledge types to create usable research for policymakers, which is designed to serve the public good from the start. This type of approach is exemplified in recent testimony to the Presidential Commission for the Study of Bioethical Issues on early integration between social and natural sciences within the Brain Initiative (Fisher, 2014). There is much work to be done, conceptually and practically, to understand how science and technology can advance the "public good" without creating unacceptable trade-offs. To this end, we suggest that normative anchors exist, e.g. human rights, distributive justice, and sustainability, to which geoengineering discourses must remain explicitly moored (von Schomberg, 2013; van den Hoven, 2013). Specifying the societal goals for those actively pursuing geoengineering research would provide an alternative evaluation scheme or a new mechanism to assess "progress". Such a scheme would help prevent geoengineering from being captured either by experts acting within an inappropriate normal science paradigm or by states or even private entities with narrow views of risks and benefits. Anticipatory governance provides a framework for embedding such normative anchors in a deliberative and inclusive process that recognises expertise – of many sorts – as not only input for, but also subject to, the collective learning that society needs to undertake to wisely govern geoengineering.

Our evaluation of five reports on geoengineering governance identifies gaps in their various approaches to foresight, engagement, and integration. There continues to be a predictive approach – we would term it a predictive fallacy – that views action as best supported through more realistic and detailed models, rather than through an exploratory approach that asks what society desires for the future, and accepts that futures are made step-by-step, rather than predicted and then achieved. Public involvement is often cast as means to educate the masses in the hope that more knowledge equates with better decisions, rather than engaging in value-based deliberation and pluralistic decision-making. Inter- and trans-disciplinary knowledge integration is challenged by the specialisation of labour between social and natural scientists and between knowledge producers and decision-makers.

We offer anticipatory governance as a vision for growing the civic capacity to guide the emergence of novel technologies. Its principles are observed in each of the five reports reviewed, and yet those high-level principles are not systematically supported by specific recommendations that contribute to building this capacity. Nonetheless, the very fact that such efforts have emerged around the governance of geoengineering research and that the principles are honoured, if sometimes in the breach, in such disparate efforts, suggests to us that a social capacity to wisely govern powerful emerging technologies may itself now be emerging. Indeed, we suspect that geoengineering's combination of uncertainty, logic and scariness are motivating an awareness among both expert and activist communities that the normal science and risk paradigms are unequal to the governance task at hand. If anticipatory governance had not already been invented, now, it appears, would be the time to do so.

First published online as a working paper in 2015.

Notes

1 Global Economic Forum (2012, p. 22) contrasts it with precaution: "More promising is the approach of 'anticipatory governance.' In this model, regulators accept the impossibility of anticipating the potential trajectory of innovations based only on past experience. They embrace the need for dynamic safeguards that can evolve with the system they are safeguarding. Anticipatory governance implies close, real-time monitoring in the direction in which innovations evolve, and involves defining safeguards flexible enough to be continually tightened or adapted in response to emerging risks and opportunities. The model of anticipatory governance is attracting attention in fields ranging from climate change to personalized medicine".
2 Consultative responses are accessible at https://royalsociety.org/policy/publications/2009/geoengineering-climate/

35

CONCLUSION

Geoengineering our climate into the future

Sean Low and Jason J. Blackstock

If this book has been an attempt to capture the evolution of geoengineering's politics since 2006, then how do we address the politics of tomorrow? It may not be useful to speculate too deeply into the future – for example, about the conditions that might undergird deployment – as there are simply too many unknown, contingent factors. However, two entwined trends can be pinpointed from the current landscape, which might be of relevance to directions taken in the near term.

First, mainstream politics have yet to seriously intrude upon geoengineering research. This is not to give short shrift to the work done by a reflective community of researchers in the natural, applied, and social sciences, in exploring the human dimensions of an engineered climate. Rather, we highlight that academic networks drive our understandings of those dimensions. This is the second point: the book you hold is a moment in time captured – and limited – by the debate's protagonists. Scientists trace histories (Part I), project, imagine, and communicate challenges (Part II), engage (or invent) incoming constituencies (Part II and V), produce driving arguments for (and against) research and experimentation, and pinpoint or propose governance systems to mitigate or enable what we see as at stake in this debate (Part IV and VI). We currently act as entrepreneurs in discourse, research, and even policy, curating the information that sets the boundaries of debate for wider audiences.

All of this is, or should be, uncomfortable territory for the geoengineering researcher. There is no consensus on how to navigate this space, only trends embedded with tensions. On one hand, the post-Crutzen debate has been informed by a militant sense of caution in technology development, has incorporated exploration of ethical, societal, and political dimensions of risk alongside the technical and climatic, and has welcomed engagement with as wide a spectrum of stakeholders as permitted by time and funding (and perhaps imagination!). Arguments that geoengineering debates are limited to the global North or have the potential to perpetuate certain dynamics of the present (for example, the global carbon economy, or North-South inequities) are welcome warnings against undesirable futures (see Chapter 27). However, they do not reflect any intent that we, as long-time participants in this debate, can see. A rich field of societal concerns and stakeholders, driven by scientists' recognition of a need to forestall technocracy, has been installed at the centre of discussions since (at least!) the publishing of the Royal Society's seminal report (2009) and the Asilomar International Conference on Climate Intervention Technologies (2010). This trend has since been reinforced by successive research programmes and consortia, commissioned reports, international conferences, and initiatives for governance scoping and stakeholder engagement – a list of which can be found following the introduction to this volume.

At the same time, research itself is changing in response to new pressures and incentives, some internal and others external. There is an increasing – though still disputed – sense that much of the low-hanging fruit has been picked in terms of regarding geoengineering as a "big picture" enterprise of managing the global climate. There is a growing shift in research from the abstract to the technology-specific and the policy-driven and a sense that new political actors and actions are needed to move the needle. Ethics, modelling, and risk assessment are becoming more fine-grained and agenda-specific, in terms of impacts, technology deployed, and regions explored (Parts I, II, III, IV, VI). Calls for disaggregating the broad umbrella term of geoengineering into its constituent suites of SRM and CDR, or even individual techniques, have become prevalent, in order to reveal specific intersections with agriculture, development, energy, security, and other challenges. An orientation towards producing knowledge for decision-support can be seen in these developments as well, alongside integrating geoengineering research more properly with concurrent strategies for mitigation, adaptation, and the global climate regime's assessment and negotiation processes in the aftermath of the 2016 Paris Agreement (Part IV). Governance proposals have evolved from an early emphasis on the downstream constraining of deployment to include the upstream enabling of explorative research (Part VI). Calls for field experimentation in SRM approaches (e.g. Dykema et al., 2014) and pilot projects for CDR in both bioenergy CCS (Smith et al., 2015) and direct air capture (Chapter 17), new high-level scoping initiatives like the Carnegie Climate Geoengineering Governance Initiative (C2G2), and the proposed fit between geoengineering approaches and the Paris Agreement's articles and targets (Craik and Burns, 2016) are all viewed with optimism as much as caution for their potential to alter debates on geoengineering – and by extension, global climate action.

How, then, does one preach and practise caution in research and technology development, while consistently expanding the spaces in and dimensions by which it is discussed? This collides with a second tension: there may come a time when the ability to frame or govern geoengineering approaches will largely move out of the hands of scientists into publics, governments, and international regimes assessing the fit between engineering the climate and their own evolving agendas – as was true of the climate change debate itself. The Convention on Biological Diversity and London Convention and Protocol have brought concerns regarding global biodiversity and the marine environment into geoengineering (Part IV); the effects of the Paris Agreement or other regimes on the debate are unfolding. What contingencies might similarly trigger attention in civil society, the private sector, or national governments, particularly outside Europe and North America (Part V)? Could the UNFCCC spur increased consideration? Could technological innovations, a rising wave of patent applications, the entry of state-sponsored programmes in technological development or even formal support of the idea of engineering the climate, or a set of national legislations regulating experimentation trigger a cascade of supporting and opposing actions? The lack of catalytic activity since Crutzen's article should not be overstated; actions in a nascent, still indeterminate landscape can radically alter political realities and future expectations.

What can researchers do to recognise and navigate the inherently political nature of research? And if the shaping influences of researchers eventually pass to new stakeholders, how can we navigate this transition with foresight and humility?

We do not dispute the general move from prohibitive to enabling governance schemes, but steps need to be taken to resolve tensions on the shape of research governance that have been ongoing for at least a half-decade. A number of influential researchers have long claimed that political imaginaries – particular ones focusing on risk – have begun to outstrip actual scientific knowledge and technology development in geoengineering (e.g. Parson and Keith, 2013). These researchers are well aware of societal implications in research and field experimentation, but there is a prioritisation of technical knowledge as a basis upon which stakeholder engagement and policy can be built. Regarding upstream governance, particularly regarding SRM field tests, researchers argue that additional and newly contrived forms of governance more burdensome than existing assessment procedures from universities, funding bodies, and existing environmental laws are stifling necessary research.

Others counter that it is politics within and external to the scientific community that shapes the priorities and stakes of research. If wider, potentially disruptive agendas are going to play a role in research, these should be accounted for at an early stage (e.g. Stilgoe et al., 2013a). Principles and procedures generated additionally by researchers, or external forms of regulation imposed from UN legal conventions, or hybrid collections thereof, may not forestall or even effectively mediate the contestation of research with such deep uncertainties (see Part IV and VI). Democratising the debate may even accelerate these tensions. Yet, they represent options in an ongoing conversation on how to increase and incorporate understandings of geoengineering's political contexts, in a way that more *laissez-faire* modes of research governance might not. It is more important to note the concept of political inclusion: we need to set the stage with early actions to forestall contestation between more nationalised, securitised agendas.

Calls for research governance to adhere to principles of anticipation or responsible innovation are in this context also welcome (Part VI). This relates to a fundamental issue in both SRM and CDR: how should we discuss or regulate technologies – and the challenges they pose – that largely do not exist, and without a consensus on whether they should exist? Indeed, the "future" is often used as a rhetorical device with which to make claims on the shape and direction of current research and politics. How should we put "facts on the ground" – technological characterisations in popular, academic, and policy literatures, and in modes of governance – in a way that is sensitive to their dependence on imaginaries, and their potential to close down particular pathways of technological framing, development, and regulation? Anticipation provides useful guidelines for how to proactively explore and manage concerns in future-oriented research, alongside regulatory mechanisms that seek to constrain the negative side effects of research – as if these are objective or obvious.

In the end, this chapter seeks to sound – or echo – a note of caution. Geoengineering approaches, if ever developed, will not emerge fully formed as a functional fleet of aircraft or ships seeding the atmosphere or a fully scaled infrastructure for capturing carbon from the air. They will emerge in a piece-meal fashion, subject to a wide range of actors, issues, and agendas in global politics at every stage in between conceptualisation and deployment, and possibly, with considerably different final shapes and challenges than those of which we presently conceive. And this behoves us to pay attention to the ever-evolving landscape within which geoengineering – not as fully fledged technologies and deployment strategies, but as immature and unscaled technologies, imaginaries, and research avenues – must uneasily grow up.

BIBLIOGRAPHY

Abbott, K.W., and D. Snidal. (2009). "The Governance Triangle: Regulatory Standards Institutions and the Shadow of the State." In Mattli, W. and Woods, N. (eds.). *The Politics of Global Regulation*. Princeton and Oxford: Princeton University Press. p. 44–88.

Achterhuis, H. (ed.). (2001). *American Philosophy of Technology: the Empirical Turn*. Robert P. Crease (trans.). Bloomington, IN: Indiana University Press.

Adger, N., Aggarwal, P., Agrawala, S., Alcamo, J., et al. (2007). *Climate Change 2007: Impacts, Adaptation and Vulnerability, Summary for Policymakers*. Geneva, Switzerland: IPCC Secretariat.

Albrecht, B.A. (1989). "Aerosols, Cloud Microphysics and Fractional Cloudiness." *Science* 245(4923): 1227–1230.

Alley, R., Marotzke, J., Nordhaus, W.D., Overpeck, J.T., Peteet, D.M., Pielke, R.A., Pierrehumbert, R.T., Rhines, P.B., Stocker, T.F., Talley, L.D., and Wallace, J.M (2003). "Abrupt Climate Change." *Science* 299(5615): 2005–2010.

Alley, R., Berntsen, T., Bindoff. N.L., Chen, Z., et al. (2007). *Climate Change 2007: The Physical Science Basis, Summary for Policymakers*. Geneva, Switzerland: IPCC Secretariat.

Alleyne, R. (2009). "Geo-engineering Should Be Developed as Insurance Against Dangerous Climate Change." *The Telegraph*. September 1. Available at: http://www.telegraph.co.uk/science/6122322/Geo-engineering-should-be-developed-as-insurance-against-dangerous-climate-change.html

Amos, J. (2010). "Key Component Contract for Iter Fusion Reactor." *BBC News*, Science and Environment. October 14. Available at: http://www.bbc.co.uk/news/science-environment-11541383

Anderson, K., and Peters, G. (2016). "The Trouble with Negative Emissions." *Science* 354(6309): 182–183.

Andersson, M. (2012). "At War over Geoengineering." *The Guardian*, Letters. February 9. Available at: http://www.theguardian.com/environment/2012/feb/09/at-war-over-geoengineering

Anselm, J., and A. Hansson. (2014a). "An Analysis of the Geoengineering Advocacy Discourse in the Public Debate." *Environmental Humanities* 5: 101–123.

Anselm, J., and A. Hansson. (2014b). "Battling Promethean Dreams and Trojan Horses: Revealing the Critical Discourses of Geoengineering." *Energy Research & Social Science* 2: 135–144.

APS (2011). Socolow, R., Desmond, M., Aines, R., Blackstock, J.J., Bolland, O., Kaarsberg, T., Lewis, N., Mazzotti, M., Pfeffer, A., Sawyer, K., Siirola, J., Smit, B., and Wilcox, J. *Direct Air Capture of CO_2 with Chemicals. A Technology Assessment Report for the American Physical Society Panel on Public Affairs.* June 1.

Araus, J.L., Slafer, G.A., Royo, C., and Serret, M.D. (2008). "Breeding for Yield Potential and Stress Adaptation in Cereals." *Critical Reviews in Plant Science* 27: 377–412.

Armeni, C. (2015a). "Global Experimentalist Governance, International Law, and Climate Change Technologies." *International and Comparative Law Quarterly* 64: 875–904.

Armeni, C. (2015b). "Geoengineering Under National Law: A Case Study of the United Kingdom." *Climate Geoengineering Governance Working Paper Series*: 023.

Armeni, C. (2015c). "Geoengineering Under National Law: A Case Study of Germany." *Climate Geoengineering Governance Working Paper Series*: 024.

Armeni, C., and Redgwell, C. (2015). "International Legal and Regulatory Issues of Climate Geoengineering Governance: Rethinking the Approach." *Climate Geoengineering Governance Working Paper Series*: 021.

Arthur, W. (1989). "Competing Technologies, Increasing Returns, and Lock-in by Historical Events." *The Economic Journal* 99: 116–131.

Asilomar. (2010). MacCracken, M., Barrett, S., Barry, R., Crutzen, P., Hamburg, S., Lampitt, R., Liverman, D., Lovejoy, T., McBean, G., Parson, E., Seidel, S., Shepherd, S., Somerville, R., and Wigley, T.M.L. *The Asilomar Conference Recommendations on Principles for Research into Climate Engineering Techniques.* Conference Report. Washington, DC: Climate Institute.

Baatz, C., Heyward, J.C., and Stelzer, H. (2016). "The Ethics of Engineering the Climate." *Environmental Values* 25(1): 1–5. doi:10.3197/096327115x14497392134766

Bacon, F. [1620] (1960). *The New Organon, and Related Writings.* Spedding, J., Ellis, R.L., and Heath, D.D. (trans.). New York: Liberal Arts Press.

Bala, G., Duffy, P.B., and Taylor, K.E. (2008). "Impact of Geoengineering Schemes on the Global Hydrological Cycle." *PNAS* 105: 7664–7669.

Banerjee, B. (2011). "The Limitations of Geoengineering Governance in a World of Uncertainty." *Stanford Journal of Law, Science, and Policy.*

Banerjee, B, Collins, G., Low, S., and Blackstock, J.J. (2013). *Scenario Planning for Solar Radiation Management.* Workshop Report. Yale Climate and Energy Institute.

Barben, D., Fisher, E., Selin, C., and Guston, D.H. (2008). "Anticipatory Governance of Nanotechnology: Foresight, Engagement, and Integration." In Hackett, E.J., Amsterdamska, O., Lynch, M.E., and Wajcman, J. (eds.). *The New Handbook of Science and Technology Studies.* Cambridge, MA: MIT Press, pp. 979–1000.

Barrett, S. (2008). "The Incredible Economics of Geoengineering." *Environmental and Resource Economics* 39: 45–54.

Barrett, S. (2014). "Solar Geoengineering's Brave New World: Thoughts on the Governance of an Unprecedented Technology." *Review of Environmental Economics and Policy* 8(2): 249–269.

Barrett, S., and A. Dannenberg. (2012). "Climate Negotiations under Scientific Uncertainty". *Proceedings of the National Academies of Sciences* 109(43): 17372–17376.

Beck, S., and Mahony, M. (2017). "The IPCC and the Politics of Anticipation." *Nature Climate Change* 7: 311–312.

Beiter, C.W., and Seidal, D.J. (2013). "A Bibliometric Analysis on Climate Engineering Research. *WIREs Climate Change* 4: 417–427.

Bellamy, R., Chilvers, J., Vaughan, N.E., and Lenton T.M. (2012). "A Review of Climate Geoengineering Appraisals. *WIREs Climate Change* 3: 597–615.

Bellamy, R., Chilvers, J., Vaughan, N.E., and Lenton, T.M. (2013). "Opening Up Geoengineering Appraisal: Multi-Criteria Mapping of Options for Tackling Climate Change." *Global Environmental Change* 23: 926–937.

Bellamy, R., Chilvers, J., and Vaughan, N.E. (2016). "Deliberative Mapping of Options for Tackling Climate Change: Citizens and Specialists 'Open up' Appraisal of Geoengineering. *Public Understanding of Science* 25: 269–286.

Bellamy, R., and Healey, P. (2018). "'Slippery Slope' or 'Uphill Struggle'? Broadening out Expert Scenarios of Climate Engineering Research and Development." *Environmental Science and Policy* 83: 1–10.

Bellamy, R., and Lezaun, J. (2017). "Crafting A Public for Geoengineering." *Public Understanding of Science* 25(4): 1–16. doi:10.1177/0963662515600965.

Benedick, R.E. (2011). "Considerations on Governance for Climate Remediation Technologies: Lessons from the "Ozone Hole." *Stanford Journal of Law, Science & Policy* 4: 6–9.

Bengtsson, L. (2006). "Geo-Engineering to Confine Climate Change: Is it At All Feasible?" *Climatic Change* 77(3–4): 229–234.

Bennett, I., and Sarewitz, D. (2006). "Too Little, Too Late? Research Policies on the Societal Implications of Nanotechnology in the United States." *Science as Culture* 15(4): 309–325.

Bereano, P.L. (1997). "Reflections of a Participant-Observer: The Technocratic/Democratic Contradiction in the Practice of Technology Assessment." *Technological Forecasting and Social Change* 54(2–3): 163–175.

Bernard, A., and Rose, Jr., W.I. (1990). "The Injection of Sulfuric Acid Aerosols in the Stratosphere by El Chichón Volcano and its Related Hazards to the International Air Traffic." *Natural Hazards* 3: 59–67.

Bernstein, M.J., Foley, R.W., and Bennett, I. (2014). "An Operationalized Post-normal Science Framework for Assisting in the Development of Complex Science Policy Solutions: The Case of Nanotechnology Governance." *Journal of Nanoparticle Research* 16: 1–14.

Beyerl, K., and Maas, A. (2014). "Perspectives on Climate Engineering from Pacific Small Island States." *Workshop Report, IASS Working Paper*, April, Potsdam.

Bickel, J. and Lane. L. (2009). "An Analysis of Climate Engineering as a Response to Climate Change." Frederiksberg, Denmark: Copenhagen Consensus Center.

Biello, D. (2012). "Pacific Ocean Hacker Speaks Out." *Scientific American*. October 24. Available at: http://www.scientificamerican.com/article/questions-and-answers-with-rogue-geoengineer-carbon-entrepreneur-russ-george/

Biermann, F., and Pattberg, P. (2008). "Global Environmental Governance: Taking Stock, Moving Forward." *Annual Review of Environment and Resources* 33: 277–294.

Bimber, B.A. (1996). *The Politics of Expertise in Congress: The Rise and Fall of the Office of Technology Assessment*. Albany, NY: State University of New York Press.

Bipartisan Policy Center (BPC). (2012). Long, J., Rademaker, J.G., Anderson, J.G., Benedick, R.E., Caldeira, K., Chaisson, J., Goldson, D., Hamburg, S., Keith, D., Lehman, R., Lowy, F., Granger, M., Sarewitz, D., Schelling, T., Shepherd, J., Victor, D.G., Whelan, D., and Winickoff, D.E. *Geoengineering: A National Strategic Plan for Research on the Potential Effectiveness, Feasibility, and Consequences of Climate Remediation Technologies*. Task Force on Climate Remediation Research, Bipartisan Policy Center.

Birnie, P.W., Boyle, A.E., and Redgwell, C. (2009). *International Law and the Environment*. Oxford: Oxford University Press.

Black, R. (2012). "Geoengineering: Risks and Benefits." *BBC*, section. Science & Environment. August 24. Available at: http://www.bbc.co.uk/news/science-environment-19371833.

Blackstock, J.J., and Ghosh, A. (2011). "Does Geoengineering Need a Global Response – and of What Kind?" *Background Paper*, Solar Radiation Management Governance Initiative, UK Royal Society, March.

Blackstock, J.J., and Long, J.C.S. (2010). "The Politics of Geoengineering." *Science* 327(5965): 527.

Bodansky, D. (1996). "May We Engineer the Climate?" *Climatic Change* 33(3): 309–321.

Bodansky, D. (2012). "What's in a Concept? Global Public Goods, International Law, and Legitimacy." *European Journal of International Law* 23(3): 651–668.

Bodansky, D. (2013). "The Who, What, and Wherefore of Geoengineering Governance." *Climatic Change* 121(3): 539–551.

Bodle, R. (2010). "Geoengineering and International Law: The Search for Common Legal Ground." *Tulsa Law Review* 46: 305–322.

Bodle, R., Oberthür, S., Donat, L., Homann, G., Sina, S., and Tedsen, E. (2014). *Options and Proposals for the International Governance of Geoengineering Climate Change.* Dessau-Roßlau, Germany: Umweltbundesamt.

Bony, S., and Dufresne, J.L. (2005). "Marine Boundary Layer Clouds at the Heart of Tropical Cloud Feedback Uncertainties in Climate Models." *Geophysial Research Letters* 32(20).

Borgmann, A. (1984). *Technology and the Character of Contemporary Life: A Philosophical Inquiry.* Chicago, IL: University of Chicago Press.

Borgmann, A. (2011). "Science, Ethics, and Technology and the Challenge of Global Warming." In Scott, D. and Francis, B. (eds.). *Debating Science: Deliberation, Values and the Common Good.* Amherst, NY: Humanities Books. pp. 169–177.

Boucher, O., Forster, P.M., Gruber, N., Ha-Duong, M., Lawrence, M.G., Lenton, T.M., Maas, A., and Vaughan, N.E. (2014). "Rethinking Climate Engineering Categorization in the Context of Climate Change Mitigation and Adaptation." *WIREs Climate Change* 5: 23–35. doi: 10.1002/wcc.261

Boucher, O. et al. (2013). "Clouds and Aerosols." In Stocker, T.F. et al. (eds.). *Climate Change 2013: The Physical Science Basis.* Cambridge: Cambridge University Press. pp. 571–658.

Boyce, D., Lewis, M., and Worm, B. (2010). "Global Phytoplankton Decline Over the Past Century." *Nature* 466: 591–596.

Boyd, O. (2012). "China Could Move First to Geoengineer the Climate." *Chinadialogue*, April 30. Available at: http://tom.chinadialogue.net/article/show/single/en/5952-China-could-move-first-to-geoengineer-the-climate

Bracmort, K., and Lattanzio, R.K. (2013). *Geoengineering: Governance and Technology Policy.* Washington, DC: Congressional Research Service.

Brand, S. (2009). *Whole Earth Discipline: Why Dense Cities, Nuclear Power, Transgenic Crops, Restored Wildlands and Geoengineering are Necessary.* New York: Penguin.

Brewer, P.G. (2007). "Evaluating a Technological Fix for Climate." *Proceedings of the National Academy of Sciences* 104(24): 9915–9916.

Brey, P. (2010). "Philosophy of Technology after the Empirical Turn." *Techne: Research in Philosophy and Technology* 14(1): 36–48.

Briggs, C. (2010). "Environmental Change, Strategic Foresight, and Impacts on Military Power." *Parameters* 40(3): 1–15.

Briggs, R. (2002). *Witches and Neighbours: The Social and Cultural Context of European Witchcraft* (Second edition). Oxford: Blackwell.

Bronk, D. (1975). "The National Science Foundation: Origins, Hopes, and Aspirations." *Science* 188(4187): 409–414.

Brossard, D., Scheufele, D.A., Kim E., and Lewenstein, B.V. (2009). "Religiosity as a Perceptual Filter: Examining Processes of Opinion Formation about Nano-technology." *Public Understanding of Science* 18(5): 546–558.

Brown, M. (2012). "First Test of Floating Volcano Geoengineering Project Cancelled." *Wired*. May 16. Available at: http://www.wired.co.uk/news/archive/2012-05/16/geoengineering-cancelled

Brown, S. (2009). "The New Deficit Model." *Nature Nanotechnology* 4(10): 609–611.

Brown Weiss, E. (1992). "Intergenerational Equity: A Legal Framework for Global Environmental Change." In *Environmental Change and International Law*. Brown Weiss, E. (ed.). Tokyo: United Nations University Press.

Brovkin, V., Petoukhov, V., Claussen, M., Bauer, E., Archer, D., and Jaeger, C. (2009). "Geoengineering Climate by Stratospheric Sulfur Injections: Earth System Vulnerability to Technological Failure." *Climatic Change* 92(3–4): 243–259.

Bryant, B.P., and Lempert, R.J. (2010). "Thinking Inside the Box: A Participatory, Computer-Assisted Approach to Scenario Discovery." *Technological Forecasting and Social Change* 77(1): 34–49.

Buck, H.J. (2013). "Climate Engineering: Spectacle, Tragedy or Solution? A Content Analysis of News Media Framing." In C. Methmann, D. Rothe, and B. Stephan (eds.). *Interpretive Approaches to Global Climate Governance. Deconstructing the Greenhouse*. New York: Routledge (Interventions), pp. 166–181.

Buck, H.J. (2015). "Climate Engineering as a Right-Wing Technology, and Other Stories." *Forum for Climate Engineering Assessment Blog*. February 15. Available at: http://dcgeoconsortium.org/2015/02/15/climate-engineering-right-wing-technology-stories-holly-buck/

Buck, H.J. (2018). "Village Science Meets Global Discourse: The Haida Salmon Restoration Corporation's Ocean Iron Fertilization Experiment." In Blackstock, J.J. and Low, S. (eds.). *Geoengineering Our Climate? Ethics, Politics and Governance*. London: Earthscan from Routledge, pp. 107–112 (this volume).

Buesseler, K.O., Doney, S.C., Karl, D.M., Boyd, P.W., Caldeira, K., Chai, F., Coale, K.H., de Baar, H.J.W., Falkowski, P.G., Johnson, K.S, Lampitt, R.S., Michaels, A.F., Naqvi, S.W.A., Smetacek, V., Takeda, S., and Watson, A.J. (2008). "Ocean Iron Fertilization— Moving Forward in a Sea of Uncertainty." *Science* 319(5860): 162–162.

Bunzl, M. (2008). "An Ethical Assessment of Geoengineering." *Bulletin of the Atomic Scientists* 64(2): 18–18.

Bunzl, M. (2009). "Researching Geoengineering: Should Not or Could Not?" *Environmental Research Letters* 4(4): 045104.

Bunzl, M. (2011). "Geoengineering Harms and Compensation." *Stanford Journal of Law, Science & Policy* 4: 69–75.

Burke, M.B., Lobell, D.B., and Guarino, L. (2009). "Shifts in African Crop Climates by 2050, and the Implications for Crop Improvement and Genetic Resources Conservation." *Global Environmental Change* 19(3): 317–325.

Burns, W.C.G. (2011). "Climate Geoengineering: Solar Radiation Management and Its Implications for Intergenerational Equity." *Stanford Journal of Law, Science & Policy* 4: 37–55.

Cairns, R. (2014a). "Climate Geoengineering: Issues of Path-dependence and Socio-technical Lock-in." *WIREs Climate Change* 5(1): 649–661. doi: 10.1002/wcc.296.

Cairns, R. (2014b). "Climates of Suspicion: 'Chemtrail' Conspiracy Narratives and the International Politics of Geoengineering." *Geographical Journal* [online]. November 25. Available at: http://onlinelibrary.wiley.com/doi/10.1111/geoj.12116/abstract

Cairns, R., and Stirling, A. (2014). "'Maintaining Planetary Systems' or 'Concentrating Global Power?' High Stakes in Contending Framings of Climate Geoengineering." *Global Environmental Change* 28: 25–38.

Caldeira, K., and Keith, D.W. (2010). "The Need for Climate Engineering Research." *Issues in Science and Technology* 27(1): 57–62.

Caldeira, K., and Wood, L. (2008). "Global and Arctic Climate Engineering: Numerical Model Studies." *Philosophical Transactions of the Royal Society A* 366: 4039–4056.

Cao, L., and Caldeira, K. (2010). "Can Ocean Iron Fertilization Mitigate Ocean Acidification?" *Climatic Change* 99(1–2): 303–311.

Cao, L., Gao, C.-C., and Zhao, L.-Y. (2015). "Geoengineering: Basic Science and Ongoing Research Efforts in China." *Advances in Climate Change Research* 6(3–4): 188–196.

Caplan, A.L. (1986). "The Ethics of Uncertainty: The Regulation of Food Safety in the United States." *Agriculture and Human Values* 3(1–2): 180–190.

Carbon Engineering. (2011). *Comments on the APS report on Direct Air Capture*. Available at: http://static.squarespace.com/static/51957744e4b088893b86e2f3/t/51b22658e4b0c62198bcd0a5/1370629720562/CE_APS_DAC_Comments.pdf

Carbon Engineering. (2013). *Direct Air Capture as an Enabler of Ultra-Low Carbon Fuels*. Available at: http://static.squarespace.com/static/51957744e4b088893b86e2f3/t/51b22bb4e4b0df9f07656a15/1370631092291/CE-DAC-CCS-Comparison.pdf

Carnegie Council for Ethics in International Affairs (2018). Carnegie Council for Ethics in International Affairs (2018) Carnegie Climate Governance Geoengineering Initiative (C2G2). Available at: https://www.c2g2.net/

Carr, W.A, Mercer, A., and Palmer, C. (2012). "Public Concerns about the Ethics of Solar Radiation Management." In Preston, C. (ed.). *Engineering the Climate: The Ethics of Solar Radiation Management*. Plymouth, UK: Lexington Books.

Carr, W.A., Preston, C.J., Yung, L., Szerszynski, B., Keith, D.W., and Mercer, A.M. (2013). "Public Engagement on Solar Radiation Management and Why It Needs to Happen Now. *Climatic Change* 121: 567–577.

Caviezel, C., and Revermann, C. (2014). *Climate Engineering*. Report No. 159. Office of Technology Assessment of the German Parliament.

CBC News (2013). "BC Village's Ocean Fertilization Experiment Probed." CBC News. 28 March. Available at: https://www.cbc.ca/news/canada/british-columbia/b-c-village-s-ocean-fertilization-experiment-probed-1.1396495

CEEW (2014). *CEEW Conference Report: Climate Geoengineering Governance*. Available at: http://ceew.in/pdf/ceew-insis-cgg-conference-report-29oct14.pdf

Central People's Government of the People's Republic of China (CPG). (2005). "Regulation of Weather Modification." Available at: http://www.gov.cn/yjgl/2005-09/27/content_70707.htm

CERN. *The Structure of CERN*. Website. Available at: http://public.web.cern.ch/public/en/About/Structure-en.html

CERN. *Member States*. Website. Available at: http://public.web.cern.ch/public/en/About/Global-en.html#observers

CGIAR *Who's Who*. Available at: http://www.cgiar.org/who/index.html

Chayes, A., and Chayes, A.H. (1995). *The New Sovereignty: Compliance with International Regulatory Agreements*. First ed. Cambridge, MA: Harvard University Press.

Chhetri, N., and Farooque, M. (2012). "World Wide Views on Biodiversity." Available at: http://archive.cspo.org/projects/wwvbio/

Chhetri, N., and Grossman, G.M. (2012). "Policy makers versus People." *The Cairo Review of Global Affairs* Spring: 118–125.

Christensen, M.W., and Stephens, G.L. (2011). "Microphysical and Macrophysical Responses of Marine Stratocumulus Polluted by Underlying Ships: Evidence of Cloud Deepening." *Journal of Geophysical Research* 116: D03201, doi: 10.1029/2010JD014638.

Christensen, M.W., and Stephens, G.L. (2012). "Microphysical and Macrophysical Responses of Marine Stratocumulus Polluted by Underlying Ships: 2. Impacts of Haze on Precipitating Clouds." *Journal of Geophysical Research* 117(D3): D11203. doi: 10.1029/2011JD017125.

Christofilos, N.C. (1959). "The Argus Experiment." *Journal of Geophysical Research* 64, 869–875.

Cicerone, R.J. (2006). "Geoengineering: Encouraging Research and Overseeing Implementation." *Climatic Change* 77(3–4): 221–226.

Clingerman, F. (2012). "Between Babel and Pelagius: Religion, Theology, and Geoengineering." In Preston, C. (ed.). *Engineering the Climate: The Ethics of Solar Radiation Management.* Plymouth, UK: Lexington Books.

Clingerman, F. (2014). "Geoengineering, Theology, and the Meaning of Being Human." *Zygon* 49(1): 6–21.

CNA Corp. (2007). *National Security and the Threat of Climate Change.* Available at: http://securityandclimate.cna.org/report/

Collingridge, D. (1980). *The Social Control of Technology.* New York: St. Martin's Press.

Connor S. (2009). "Man-made eruptions – 'Plan B' in the Battle for the Planet." *The Independent.* September 2. Available at: http://www.independent.co.uk/environment/climate-change/manmade-eruptions-ndash-plan-b-in-the-battle-for-the-planet-1780268.html

Convention on Biological Diversity (CBD). (1992). *Convention on Biological Diversity.* Available at: https://www.cbd.int/doc/legal/cbd-en.pdf

Convention on Biological Diversity (CBD). (2008). Ninth Meeting of the Conference of Parties to the Convention on Biological Diversity, May 19–30, Decision IX/16—Biodiversity and Climate Change, UN Doc. UNEP/CBD/COP/DEC/IX/16.

Convention on Biological Diversity (CBD). (2010a). Tenth Meeting of the Conference of Parties to the Convention on Biological Diversity, Oct. 18–29, Decision X/33—Biodiversity and Climate Change, UN Doc. UNEP/CBD/COP/DEC/X/33.

Convention on Biological Diversity (CBD). (2010b). "Climate Related Geoengineering and Diversity." Available at: http://www.cbd.int/climate/geoengineering

Convention on Biological Diversity (CBD). (2012). Eleventh Meeting of the Conference of Parties to the Convention on Biological Diversity, Oct. 8–19, Decision XI/20—Climate-related Geoengineering, UN Doc. UNEP/CBD/COP/DEC/XI/20.

Convention on Biological Diversity (CBD). (2016). Thirteenth Meeting of the Conference of Parties to the Convention on Biological Diversity, Dec. 4–17, Decision XIII/14—Climate-related Geoengineering, UN Doc. UNEP/CBD/COP/DEC/XIII/14.

Cook-Deegan, R. (1994). *The Gene Wars: Science, Politics, and the Human Genome.* New York: W.W. Norton.

Cooper, R.N. (2010). "The Case for Charges on Greenhouse Gas Emissions." In Aldy, J.E. and Stavins, R.N. (eds.) *Post-Kyoto International Climate Policy: Implementing Architectures for Agreement.* Cambridge: Cambridge University Press.

Corner, A., Parkhill, K., and Pidgeon, N. (2011). "Experiment Earth? Reflections on a Public Dialogue on Geoengineering." *Understanding Risk Working Paper* 11–02. School of Psychology, Cardiff University.

Corner, A., Parkhill, K., Pidgeon, N., and Vaughan, N.E. (2013). "Messing With Nature? Exploring Public Perceptions of Geoengineering in the UK." *Global Environmental Change* 23: 938–947.

Corner, A., and Pidgeon, N. (2010). "Geoengineering the Climate: The Social and Ethical Implications." *Environment: Science and Policy for Sustainable Development* 52(1): 24–37.

Corner, A., and Pidgeon, N. (2014). "Like Artificial Trees? The Effect of Framing by Natural Analogy on Public Perceptions of Geoengineering." *Climatic Change.* doi: 10.1007/s10584-014-1148-6.

Corner, A., Pidgeon, N., and Parkhill, K. (2012). "Perceptions of Geoengineering: Public Attitudes, Stakeholder Perspectives, and the Challenge of 'Upstream' Engagement." *WIREs Climate Change* 3(5): 451–466.

Council of the Haida Nation (2012). Joint Statement. Available at: http://www.etcgroup. org/sites/www.etcgroup.org/files/Statement%20from%20Council%20of%20 Haida%20Nations.pdf

Craik, A.N., and Burns, W.C.G. (2016). "Climate Engineering under the Paris Agreement: A Legal and Policy Primer." Special Report. Centre for International Governance Innovation.

Cressey, D. (2012). "Cancelled Project Spurs Debate over Geoengineering Patents." *Nature.* May 23. Available at: http://www.nature.com/news/cancelled-project-spurs-debate-over-geoengineering-patents-1.10690

Crookes, W. (1898). "British Association for the Advancement of Science, Bristol 1898, Inaugural Address of the President." *The Chemical News* 73(2024): 125–136.

Crutzen, P.J. (2006). "Albedo Enhancement by Stratospheric Sulfur Injections: A Contribution to Resolve a Policy Dilemma?" *Climatic Change* 77(3–4): 211–219.

Curry, C.L., Sillmann, J., Bronaugh, D., Alterskjaer, K., Cole, J.N.S., Ji, D.Y., Kravitz, B., Kristjansson, J.E., Moore, J.C., Muri, H., Niemeier, U., Robock, A., Tilmes, S., and Y, S.T. (2014). "A Multimodel Examination of Climate Extremes in an Idealized Geoengineering Experiment." *Journal of Geophysical Research: Atmospheres* 119: 3900–3923.

Dakos, V. et al. (2008). "Slowing Down as an Early Warning Signal for Abrupt Climate Change." *Proceedings of the National Academies of Sciences* 105(38): 14308–14312.

D'Alessandro, D.M., Smit, B., and Long, J.R. (2010). "Carbon Dioxide Capture: Prospects for New Materials." *Angew. Chem. Int. Ed.* 49(35): 6058–6082.

Dalton, G.F. (1970). "The Ritual Killing of the Irish Kings." *Folklore* 81(1): 1–21.

Daniels, N., and Sabin, J.E. (1997). "Limits to Healthcare: Fair Procedures, Democratic Deliberation, and the Legitimacy Problem for Insurers." *Philosophy and Public Affairs* 26: 303–350.

Denman, K. (2008). "Climate Change, Ocean Processes and Ocean Iron Fertilization." *Marine Ecology Progress Series* 364: 219–225.

Disney, J. (2013). Written interview by Buck, H.J., August 26.

Doney, S.C., Fabry, V.J., Feely, R.A., and Kleypas, J.A. (2009). "Ocean Acidification: The Other CO_2 Problem." *Annual Review of Marine Science* 1: 169–192.

Donner, S. (2007). "Domain of the Gods: An Editorial Essay." *Climatic Change* 85(3–4): 231–236.

Dryzek, J.S., and Niemeyer, S. (2008). "Discursive Representation." *American Political Science Review* 102: 481–493.

Dunoff, J.L., and Pollack, M.A. (eds.) (2012). *Interdisciplinary Perspectives on International Law and International Relations: The State of the Art.* Cambridge: Cambridge University Press.

Dupuy, J.-P. (2007). "Complexity and Uncertainty: A Prudential Approach to Nanotechnology." In Allhoff, F., Lin, P., Moor, J., and Weckert, J. (eds.). *Nanoethics: The Ethical and Social Implications of Nanotechnology.* Hoboken, NJ: John Wiley & Sons, pp. 119–131.

Dykema, J., Keith, D.W., Anderson, J.G., and Weisenstein, D. (2014). "Stratospheric Controlled Perturbation Experiment (SCoPEx): A Small-scale Experiment to Improve Understanding of the Risks of Solar Geoengineering." *Philosophical Transactions of the Royal Society A*, 372(2031): 1–21.

Dyson, F.J. (2008). "The Question of Global Warming." *The New York Review of Books*, June 12.

Dyson, F.J. (1977). Can We Control the Carbon Dioxide in the Atmosphere? *Energy* 2: 287–291.

Dyson, F.J. (1999). *The Sun, the Genome, & the Internet*. Oxford: Oxford University Press.

Easlea, B. (1980). *Witch Hunting, Magic, and the New Philosophy: An Introduction to Debates of the Scientific Revolution, 1450–1750*. Brighton, UK: Harvester Press.

The Economist. (2010). "Geoengineering: Lift Off." November 4.

Edney, K., and Symons, J. (2013). "China and the Blunt Temptations of Geoengineering: The Role of Solar Radiation Management in China's Strategic Response to Climate Change." *The Pacific Review* 27(3): 307–332. Available at: http://www.tandfonline.com/eprint/m7PvHD8PDrb9iGaKtbz3/full#.Ut7GK_Y1ioQ

Eisenberger, P.M., Cohen, R.W., Chichilnisky, G., Eisenberger, N.M., Chance, R.R., and Jones, C.W. (2009). "Global Warming and Carbon-Negative Technology: Prospects for a Lower-Cost Route to a Lower-Risk Atmosphere." *Energy & Environment* 20(6): 973–984.

Elblaus, T. (2014). *Climate Change in Four Dimensions*. Unpublished report, on file with the author (Reynolds, J.L.).

Eliseev, A.V., Chernokulsky, A.V., Karpenko, A.A., and Mokhov, I.I. (2010). "Global Warming Mitigation by Sulphur Loading in the Stratosphere: Dependence of Required Emissions on Allowable Residual Warming Rate." *Theoretical Applied Climatology*, 101(1–2): 67–81.

Erisman, J.W., Sutton, M.A., Galloway, J., Klimont, Z, and Winiwarter, W. (2008). "How a Century of Ammonia Synthesis Changed the World." *Nature Geoscience* 1: 636–639.

Espy, J.P. (1841). *The Philosophy of Storms*. Boston, MA: Little, Brown.

ETC Group. (2010a). "Open Letter Opposing Asilomar Geoengineering Conference." Available at http://www.etcgroup.org/en/node/5080

ETC Group. (2010b). "Geopiracy: The Case Against Geoengineering." ETC Group Communiqué 103, November. Available at: http://www.etcgroup.org/en/node/5217

ETC Group. (2010c). "What does the UN Moratorium on Geoengineering mean?" November 11. Available at: http://www.etcgroup.org/content/what-does-un-moratorium-geoengineering-mean

ETC Group. (2011). "Say No to the Trojan Hose." Available at: http://www.etcgroup.org/content/say-no-trojan-hose

ETC Group. (2012). "Informational Backgrounder on the 2012 Haida Gwaii Iron Dump." Accessed Sept 2013 at: http://www.etcgroup.org/content/informational-backgrounder-2012-haida-gwaii-iron-dump

ETC Group. (2018). "Opposition to Geoengineering: There's No Place like H.O.M.E." In Blackstock, J.J. and Low, S. (eds.). *Geoengineering Our Climate? Ethics, Politics and Governance*. London: Routledge, pp. 174–177 (this volume).

Ethics of Geoengineering Online Resource Center, University of Montana. Ethics: Ethics Resources. Available at: http://www.umt.edu/ethics/resourcecenter/Bibliography/ethics.php

European Commission. (2008). *Strategic Action Plan for Implementation of European Regional Repositories: Stage 2. Work Package: Public and Political Attitudes*. European Commission under the Euratom Research and Training Programme on Nuclear Energy within the Sixth Framework Programme (2002–2006) and the Swiss Federal Office for Education and Science. Available at: ftp://ftp.cordis.europa.eu/pub/fp6-euratom/docs/sapierr-2-5-public-and-political-attitudes_en.pdf

European Organization for Nuclear Research. (2008). *General Conditions Applicable to Experiments at CERN.* Available at: http://committees.web.cern.ch/committees/GeneralConditions.pdf Accessed Feb 2014.

EuTRACE (2015). Schäfer, S., Lawrence, M., Stelzer, H., Born, W., and Low, S. (eds.) *The European Transdisciplinary Assessment of Climate Engineering (EuTRACE): Removing Greenhouse Gases from the Atmosphere and Reflecting Sunlight away from Earth.* Funded by the European Union's Seventh Framework Programme under Grant Agreement 306993.

Fabry, V.J., Seibel, B.A., Feely, R.A., and Orr, J.C. (2008). "Impacts of Ocean Acidification on Marine Fauna and Ecosystem Processes." *ICES Journal of Marine Science: Journal du Conseil* 65(3): 414–432.

Fagin, D. (1992). "Tinkering with the Environment." *Newsday.* April 13. p. 7.

Fargione, J., Hill, J., Tilman, D., Polasky, S., and Hawthorne, P. (2008). "Land Clearing and the Biofuel Carbon Debt." *Science* 319(5867): 1235–1238.

Farrell, P. (2012). "6 Ways Billionaires Try to Play God on Climate." *MarketWatch.* November 2. Available at: http://www.marketwatch.com/story/6-ways-billionaires-try-to-play-god-on-climate-2012-11-02.

Fiorino, D.J. (1990). "Citizen Participation and Environmental Risk: A Survey of Institutional Mechanisms." *Science, Technology and Human Values* 15: 226–243.

Fisher, E. (2014). "Transcript, Meeting 16 Opening Remarks and Session 6." Presidential Commission for the Study of Bioethical Issues. Washington, DC. February 11.

Fleming, J.R. (2006). "The Pathological History of Weather and Climate Modification: Three Cycles of Promise and Hype." *Historical Studies in the Physical Sciences* 37: 3–25.

Fleming, J.R. (2007). "The Climate Engineers." *The Wilson Quarterly* 31(2): 46–60.

Fleming, J.R. (2010). *Fixing the Sky: The Checkered History of Weather and Climate Control.* New York: Columbia University Press.

Fleming, J.R. (2011). "Iowa Enters the Space Age: James Van Allen, Earth's Radiation Belts, and Experiments to Disrupt Them." *Annals of Iowa* 70: 301–324.

Fleming, J.R. (2018). "A History of Weather and Climate Control." In Blackstock, J.J. and Low, S. (eds.). *Geoengineering Our Climate? Ethics, Politics and Governance.* London: Earthscan from Routledge, pp. 30–33 (this volume).

Forum for Climate Engineering Assessment. (2015). "Civil Society Statements on Release of NAS 'Climate Intervention' Reports." *Forum for Climate Engineering Assessment Blog.* February 10. Available at http://dcgeoconsortium.org/2015/02/10/civil-society-statements-on-the-release-of-nas-climate-intervention-reports/

Fountain, H. (2012). "A Rogue Climate Experiment Outrages Scientists." *New York Times.* October 18. Available at: http://www.nytimes.com/2012/10/19/science/earth/iron-dumping-experiment-in-pacific-alarms-marine-experts.html?_r=0

Funtowicz, S., and Ravetz, J. (1993). "Science for the Post-Normal Age." *Futures* 25(7): 735–755.

Fox, T.A., and Chapman, L. (2011). "Engineering Geo-engineering." *Meteorological Applications* 18: 1–8.

Frankfort, H., Frankfort, H.A., Wilson, J.A., and Jacobsen, T. (1949). *Before Philosophy: The Intellectual Adventure of Ancient Man.* London: Penguin.

Galarraga, M., and Szerszynski, B. (2012). "Making Climates: Solar Radiation Management and the Ethics of Fabrication." In Preston, C. (ed.). *Engineering the*

Climate: The Ethics of Solar Radiation Management. Lexington, MA: Lexington, pp. 211–225.

Galaz, V. (2012). "Geo-engineering, Governance, and Social-Ecological Systems: Critical Issues and Joint Research Needs." *Ecology and Society* (17)1: 24.

Gardiner, S. (2009). "Is 'Arming the Future' with Geoengineering Really the Lesser Evil?" In Gardiner S., et al. (eds.). *Climate Ethics Essential Readings.* Oxford: Oxford University Press.

Gardiner, S. (2011). "Some Early Ethics of Geoengineering the Climate: A Commentary on the Values of the Royal Society Report." *Environmental Values* 20: 163–188.

Gaskell, G., Einsiedel, E., Hallman, W., Priest, S.H., Jackson, J., and Olsthoorn, J. (2005). "Social Values and the Governance of Science." *Science* 310(5756): 1908–1909.

Geden, O., and Beck, S. (2014). "Renegotiating the Global Climate Stabilization Target." *Nature Climate Change* 4: 747–748.

Geoengineering Model Intercomparison Project. *GeoMIP Publications.* Website. Available at: http://climate.envsci.rutgers.edu/GeoMIP/publications.html

Geoengineering: Parts I, II, and III. (2010). Hearing before the Committee on Science and Technology, House of Representatives, 111th Congress, 1. Available at: http://www.gpo.gov/fdsys/pkg/CHRG-111hhrg53007/pdf/CHRG-111hhrg53007.pdf

Gerard, D., and Wilson, E.J. (2008). "Environmental Bonds and the Challenge of Long-Term Carbon Sequestration." *Journal of Environmental Management* 1097, 1100.

GGF 2025 Geoengineering Governance Working Group. (2015). *Human Intervention in the Earth's Climate: The Governance of Geoengineering in 2025+.* Global Governance Futures, Robert Bosch Foundation Multilateral Dialogues.

Ghosh, A. (2010). "Harnessing the Power Shift: Governance Options for International Climate Financing. *Oxfam Research Report.* October 6.

Ghosh, A. (2011a). "International Cooperation and the Governance of Geoengineering." Keynote Lecture to the Expert Meeting on Geoengineering, Intergovernmental Panel on Climate Change, Lima. June 21. Available at: http://ceew.in/pdf/AG_International_Cooperation_IPCC_21Jun11.pdf.

Ghosh, A. (2011b). "Seeking Coherence in Complexity?: The Governance of Energy by Trade and Investment Institutions." *Global Policy* 2(Special Issue): 106–119.

Ghosh, A. (2018). "Environmental Institutions, International Research Programmes, and Lessons for Geoengineering Research." In Blackstock, J.J. and Low, S. (eds.). *Geoengineering Our Climate? Ethics, Politics and Governance.* London: Earthscan from Routledge, pp. 199–213 (this volume).

Ghosh, A., and Woods, N. (2009). "Governing Climate Change: Lessons from Other Governance Regimes." In Helm, D. and Hepburn, C. (eds.). *The Economics and Politics of Climate Change.* Oxford: Oxford University Press.

Gingrich, N. (2008). "Stop the Green Pig: Defeat the Boxer-Warner-Lieberman Green Pork Bill Capping American Jobs and Trading America's Future." *Human Events.* June 3. Available at: http://humanevents.com/2008/06/03/stop-the-green-pig-defeat-the-boxerwarnerlieberman-green-pork-bill-capping-american-jobs-and-trading-americas-future/

Glenn, J.C., and Gordon, T.J. (eds.) (2009). *Futures Research Methodology Version 3.0.* The Millennium Project.

Global Economic Forum. (2012). *Global Risks Report, 2012, seventh edition.* Global Economic Forum: Geneva.

Goeschl, T., Heyen, D., and Moreno-Cruz., J. (2013). "The Intergenerational Transfer of Solar Radiation Management Capabilities and Atmospheric Carbon Stocks." *Environmental and Resource Economics* 56(1): 85–104. doi:10.1007/s10640-013-9647-x.

Goes, M., Tuana, N., and Keller, K. (2011). "The Economics (or Lack Thereof) of Aerosol Geoengineering." *Climatic Change* 109(3–4): 719–744.

Goldenberg, S. (2014). "Al Gore says Use of Geo-engineering to Head Off Climate Disaster is Insane." *The Guardian*. January 15. Available at: http://www.theguardian.com/world/climate-consensus-97-per-cent/2014/jan/15/geo-al-gore-engineering-climate-disaster-instant-solutio

González, A.J. (2011). "The Argentine Approach to Radiation Safety: Its Ethical Basis." *Science and Technology of Nuclear Installations* 2011(910718): 1–15.

Goodell, J. (2010). *How to Cool the Planet: Geoengineering and the Audacious Quest to Fix Earth's Climate.* Boston, MA: Houghton Mifflin Harcourt.

Gordon, B. (2010). *Engineering the Climate: Research Needs and Strategies for International Coordination.* Committee on Science and Technology, House of Representatives, United States Congress, Washington, DC.

Gorman, M.E. (2012). A Framework for Anticipatory Governance and Adaptive Management of Synthetic Biology. *International Journal of Social Ecology and Sustainable Development* 3(2): 64–68.

Govindasamy, B., Caldeira, K., and Duffy, P.B. (2003). "Geoengineering Earth's Radiation Balance to Mitigate Climate Change from a Quadrupling of CO_2." *Global Planetary Change*, 37(1–2): 157–168.

Grasso, M. (2007). "A Normative Ethical Framework in Climate Change." *Climatic Change* 81(3–4): 223–246.

Gregory, J.M., Huybrechts, P., and Raper, S.C.B. (2004). "Threatened Loss of the Greenland Ice-sheet." *Nature* 428: 616–616.

Groves, D.G., and Lempert, R.J. (2007). "A New Analytic Method for Finding Policy-Relevant Scenarios." *Global Environmental Change* 17(1): 73–85.

Guillemot, H. (2017). "The Necessary and Inaccessible 1.5C Objective: A Turning Point in the Relations between Climate Science and Politics?" In Aykut, S.C., Foyer, J., and Morena, E. (eds.). *Globalising the Climate: COP21 and the Climatisation of Global Debates.* New York: Earthscan from Routledge.

Guston, D.H. (2008). "Innovation Policy: Not Just a Jumbo Shrimp." *Nature* 454: 940–941.

Guston, D.H. (2014). "Understanding Anticipatory Governance." *Social Studies of Science* 44(2): 218–242.

Guston, D.H., and Sarewitz, D. (2002). "Real-Time Technology Assessment." *Technology in Society* 24: 93–109.

Guzman, A.T. (2008). *How International Law Works: A Rational Choice Theory.* Oxford: Oxford University Press.

Haasnoot, M., Kwakkel, J.H., Walker, W.E., and ter Maat, J. (2013). "Dynamic Adaptive Policy Pathways: A Method for Crafting Robust Decisions for a Deeply Uncertain World." *Global Environmental Change* 23: 485–498.

Haida Laas. (2012). Website. June. Available at: http://www.haidanation.ca/Pages/haida_laas/hl_archives.html

Hale, B., and Dilling, L. (2011). "Geoengineering, Ocean Fertilization, and the Problem of Permissible Pollution." *Science, Technology, & Human Values* 36(2): 190–212.

Hale, E. (2012). "Geoengineering Experiment Cancelled due to Perceived Conflict of Interest." *The Guardian.* May 16. Available at: http://www.theguardian.com/environment/2012/may/16/geoengineering-experiment-cancelled

Hall, J.W., Lempert, R.J., Keller, K., Hackbarth, A., Mijere, C., and McInerney, D.J. (2012). "Robust Climate Policies Under Uncertainty: A Comparison of Robust Decision Making and Info-Gap Methods." *Risk Analysis* 32(10): 1657–1672.

Hamilton, C. (2011). "The Ethical Foundations of Climate Engineering." Available at: http://www.clivehamilton.net.au/cms/index.php?page=articles

Hamilton, C. (2013a). *Earthmasters: The Dawn of the Age of Geoengineering.* New Haven, CT: Yale University Press.

Hamilton, C. (2013b). "Why Geoengineering has Immediate Appeal to China." *The Guardian.* March 22. Available at: http://www.theguardian.com/environment/2013/mar/22/geoengineering-china-climate-change

Hanafi, A., and Hamburg, S.P. (2018). "The Solar Radiation Management Governance Initiative: Advancing the International Governance of Geoengineering Research." In Blackstock, J.J. and Low, S. (eds.) *Geoengineering Our Climate? Ethics, Politics and Governance.* London: Earthscan from Routledge, pp. 214–217 (this volume).

Hanks, C. (ed.) (2007). *Technology and Human Values.* Oxford: Wiley-Blackwell.

Hansen, J., Sato, M., Ruedy, R., Lo, K., Lea, D.W., and Medina-Elizade, M. (2006). "Global Temperature Change." *Proceedings of the National Academy of Sciences* 103(39): 14288–14293.

Hansson, A. (2012). "Colonizing the Future: The Case of CCS." In Markusson, N., Shackley, S., and Evar, B. (eds.). *The Social Dynamics of Carbon Capture and Storage: Understanding Representation, Governance and Innovation.* London: Routledge. pp. 74–90.

Harnisch, S., Uther, S., and Boettcher, M. (2015). "From 'Go Slow' to Gung Ho'? Climate Engineering Discourses in the UK, the US, and Germany. *Global Environmental Politics* 15(2): 57–78.

Harris, P. (2010). *World Ethics and Climate Change.* Edinburgh: Edinburgh University Press.

Harvard Model United Nations General Assembly. (2011). Topic Area Summaries. Available at: http://www.harvardmun.org/wp-content/uploads/2010/06/GA-TAS-2011.pdf

Haywood, J., Jones, A., Bellouin, N., and Stephenson, D. (2013). "Asymmetric Forcing from Stratospheric Aerosols Impacts on Sahelian Rainfall." *Nature Climate Change* 3(7): 660–665.

Heckendorn, P., Weisenstein, D., Fueglistaler, S., Luo, B.P., Rozanov, E., Schraner, M., Thomason, L.W., and Peter, T. (2009). "The Impact of Geoengineering Aerosols on Stratospheric Temperature and Ozone." *Environmental Research Letters* 4(4): 045108. doi:10.1088/1748-9326/4/4/045108.

Hegerl, G.C., and Solomon, S. (2009). "Risks of Climate Engineering." *Science* 325(5943): 955–956.

Held, H., and Kleinen, T. (2004). "Detection of Climate System Bifurcations by Degenerate Fingerprinting". *Geophysical Research Letters* 31: L23207.

Hester, T. (2011). "Remaking the World in Order to Save It: Applying US Environmental Laws." *Ecology Law Quarterly* 38(4): 851–892.

Hester, T. (2013). "A Matter of Scale: Regional Climate Engineering and the Shortfalls of Multinational Governance." *Carbon and Climate Law Review* 7(3): 168–176.

Heyward, C. (2013). "Situating and Abandoning Geoengineering: A Typology of Five Responses to Dangerous Climate Change." *Political Science & Politics* 46: 23–27. doi: 10.1017/S104909651200143

Honegger, M., Sugathapala, K., and Michaelowa, A. (2013). "Tackling Climate Change: Where Can the Generic Framework be Located?" *Climate Change Law Review* 2: 125–135.

Horsley, R.A. (1979). "Further Reflections on Witchcraft and European Folk Religion." *History of Religions* 19(1): 71–95.

Horton, J.B. (2012a). "OIF Accusations Fly at CBD COP 11." *Geoengineering Politics*. Weblog. October 17. Available at: http://geoengineeringpolitics.blogspot.de/2012/10/oif-accusations-fly-at-cbd-cop11.html

Horton, J.B. (2012b). "Nothing New Emerges from CBD COP 11." *Geoengineering Politics*. Weblog. October 22. Available at: http://geoengineeringpolitics.blogspot.de/2012/10/nothing-new-emerges-from-cbd-cop11.html

Horton, J.B., Parker, A., and Keith, D. (2018). "Solar Geoengineering and the Problem of Liability." In Blackstock, J.J. and Low, S. (eds.) *Geoengineering Our Climate? Ethics, Politics and Governance*. London: Earthscan from Routledge, pp. 142–146 (this volume).

House, K.Z., Baclig, A.C., Ranjan, M., van Nierop, E.A., Wilcox, J., and Herzog, H.J. (2011). Economic and Energetic Analysis of Capturing CO_2 from Ambient Air. *Proceedings of the National Academy of Science* 108(51): 20428–20433.

House of Commons, UK (HOC) (2010). *The Regulation of Geoengineering*. London: House of Commons Science and Technology Committee. London: The Stationery Office. Available at: http://www.publications.parliament.uk/pa/cm200910/cmselect/cmsctech/221/22102.htm

Hulme, M., and Mahony, M. (2010). "Climate Change: What Do We Know about the IPCC?" *Physical Geography* 34(5): 705–718.

Hume, M. (2012). "Ocean Fertilization Experiment Alarms Marine Scientists." *The Globe and Mail*. October 19. Available at: http://www.theglobeandmail.com/news/national/ocean-fertilization-experiment-alarms-marine-scientists/article4625695/

Hume, M. (2013). "Pink Salmon Reaching Fraser River in Massive Numbers." *The Globe and Mail*. September 13. Available at: http://www.theglobeandmail.com/news/british-columbia/article14298697.ece

Huttunen, S., and Hilden, M. (2013). "Framing the Controversial: Geoengineering in Academic Literature." *Science Communication* (online version) 36(1): 1–27.

Huttunen, S., Skyten, E., and Hilden, M. (2014). "Emerging Policy Perspectives on Geoengineering: An International Comparison." *The Anthropocene Review*: 1–19.

Hwang, Y.T., Frierson, D., and Kang, S. (2013). "Anthropogenic Sulfate Aerosol and the Southward Shift of Tropical Precipitation in the Late 20th Century." *Geophysical Research Letters* 40(11): 2845–2850. doi:10.1002/grl.50502.

IGBP, IOC, SCOR. (2013). "Ocean Acidification Summary for Policymakers – Third Symposium on the Ocean in a High-CO_2 World." *International Geosphere-Biosphere Programme*. Stockholm, Sweden.

Ihde, D. (2003). "Has Philosophy of Technology Arrived? A State-of-the-Art Review." *Philosophy of Science* 71: 117–131.

IISD Reporting Service. (2010). "Summary of the 10th Conference of the Parties to the CBD, October 18–29." *Earth Negotiations Bulletin* 9(544): 1–30. Available at: http://www.iisd.ca/vol09/enb09544e.html

Implementing Geological Disposal of Radioactive Waste Technology Platform (IGD-TP) (2012). "Deployment Plan 2011–2016". Funded by European Community's Seventh Framework Programme (FP7/2007-20013) under grant agreement n° 249396.

Institute of Medicine, National Academy of Sciences, and National Academy of Engineering. (1992). *Policy Implications of Greenhouse Warming*. Washington, DC: National Academies Press.

Intergovernmental Panel on Climate Change (IPCC). (1990). Policymakers Summary of the Response Strategies Working Group of the Intergovernmental Panel on Climate Change (Working Group III). In Bernthal, F.M., and Secretariat (eds.) *Digitized by the Digitization and Microform Unit, UNOG Library, 2010*. Available at: https://

www.ipcc.ch/publications_and_data/publications_ipcc_first_assessment_1990_wg3.shtml

Intergovernmental Panel on Climate Change (IPCC). (1995). Summary for Policymakers: Scientific-Technical Analyses of Impacts, Adaptations, and Mitigation of Climate Change. A Report of Working Group II of the Intergovernmental Panel on Climate Change. In Watson, R.T. and the Core Writing Team (eds.). Cambridge and New York: Cambridge University Press.

Intergovernmental Panel on Climate Change (IPCC). (2001). Summary for Policymakers. In: Climate Change 2001: Mitigation. A Report of Working Group III of the Intergovernmental Panel on Climate Change. In Watson, R.T. and the Core Writing Team (eds.). Cambridge and New York: Cambridge University Press.

Intergovernmental Panel on Climate Change (IPCC). (2005). *IPCC: Carbon Capture and Storage*. Cambridge: Cambridge University Press.

Intergovernmental Panel on Climate Change (IPCC). (2011). Boucher, O., Gruber, N., and Blackstock, J.J., *Summary of the Synthesis Session In: IPCC Expert Meeting Report on Geoengineering.* [Edenhofer, O., Field, C., Pichs-Madruga, R., Sokona, Y., Stocker, T., Barros, V., Dahe, Q., Minx, J., Mach, K., Plattner, G.-K., Schlömer, S., Hansen, G., and Mastrandrea, M. (eds.).] IPCC Working Group III Technical Support Unit, Potsdam Institute for Climate Impact Research, Potsdam, Germany. June. Available at: https://www.ipcc-wg1.unibe.ch/publications/supportingmaterial/EM_GeoE_Meeting_Report_final.pdf

Intergovernmental Panel on Climate Change. (2012). Special Report on Managing the Risks of Extreme Events and Disasters to Advance Climate Change Adaptation. Summary for Policy Makers. Available at: http://www.ipcc-wg2.gov/SREX/

Intergovernmental Panel on Climate Change. (2013). *Climate Change 2013 The Physical Science Basis: Summary for Policymakers.* Working Group I Contribution to the Firth Assessment Report of the Intergovernmental Panel on Climate Change. Available at: http://www.climatechange2013.org/images/uploads/WGI_AR5_SPM_brochure.pdf

Intergovernmental Panel on Climate Change (IPCC). (2014). Climate Change 2014: Mitigation of Climate Change. *Contribution of Working Group III to the Fifth Assessment Report of the Intergovernmental Panel on Climate Change.* In Edenhofer, O., Pichs-Madruga, R., Sokona, Y., Farahani, E., Kadner, S., Seyboth, K., Adler, A. , Baum, I., Brunner, S., Eickemeier, P., Kriemann, B., Savolainen, J., Schlömer, S., von Stechow, C., Zwickel, T., and Minx, J.C. (eds.). Cambridge and New York: Cambridge University Press.

International Committee of the Red Cross. (ICRC). (n.d.) "Convention on the Prohibition of Military or Any Hostile Use of Environmental Modification Techniques, December 10, 1976." Available at: http://www.icrc.org/ihl.nsf/INTRO/460

International Energy Agency. (2012). CO_2 *Emissions from Fuel Combustion: Highlights.* Available at: http://www.iea.org/co2highlights/co2highlights.pdf

International Law Commission. (2001a). "Draft Articles on Prevention of Transboundary Harm from Hazardous Activities." In *Report of the International Law Commission, 53rd Session, Official Records of the General Assembly.* UN Doc A/56/10.

International Law Commission. (2001b). "Draft Articles on Responsibility of States for Internationally Wrongful Acts." In *Report of the International Law Commission, 53rd Session, Official Records of the General Assembly*; UN Doc. A/56/10.

International Law Commission. (2006). "Draft Principles on the Allocation of Loss in the Case of Transboundary Harm Arising out of Hazardous Activities." In *Report of the International Law Commission, 58th Session, Official Records of the General Assembly.* UN Doc. A/61/10.

International Maritime Organization (IMO). (2013). "Marine Geoengineering including Ocean Fertilization to Be Regulated under Amendments to International Treaty." Press Briefing. October 18. Available at: http://www.imo.org/MediaCentre/PressBriefings/Pages/45-marine-geoengieneering.aspx#.VM-4DSeUf_4

IOPC Funds. (2011). *Annual Report*. London: IOPC Funds.

IPSOS Mori. (2010). *Experiment Earth? Report on a Public Dialogue on Geoengineering*. Report to the Natural Environment Research Council (NERC).

Irvine, P.J., Kravitz, B., Lawrence, M.G., and Muri, H. (2016). "An Overview of the Earth System Science of Solar Geoengineering." *WIREs Climate Change* 7(6): 815–833.

Irvine, P.J., Ridgwell, A., and Lunt, D.J. (2010). "Assessing the Regional Disparities in Geoengineering Impacts." *Geophysical Research Letters* 37(18): 1–6.

Irvine, P.J., Sriver, R., and Keller, K. (2012). "Strong Tension Between the Objectives to Reduce Sea-level Rise and Rates of Temperature Change through Solar Radiation Management." *Nature Climate Change* 2: 97–100.

Izrael, Y.A., et al. (2009). "Field Experiment on Studying Solar Radiation Passing Through Aerosol Layers." *Russian Meteorology and Hydrology* 34(5): 265–273.

Izrael, Y.A., Zakharov, V.M., Petrov, N.N., Ryaboshapko, A.G., Ivanov, V.N., Savchenko, A.V., Andreev, Yu V., et al. (2010)."Field Studies of a Geo-Engineering Method of Maintaining a Modern Climate with Aerosol Particles." *Russian Meteorology and Hydrology* 34(10): 635–638.

James, W. (1981). *Pragmatism*. Bruce Kuklick (ed.). Indianapolis, IN: Hackett.

Jamieson, D. (1996). "Intentional Climate Change." *Climatic Change* 33: 323–336.

Jamieson, D. (2007). "The Moral and Political Challenges of Climate Change." In Moser, S.C. and Dilling, L. (eds.). *Creating a Climate for Change: Communicating Climate Change and Facilitating Social Change*. New York: Cambridge University Press, pp. 475–482.

Jamieson, D. (2010). "Climate Change, Responsibility and Justice." *Science and Engineering Ethics* 16(3): 431–445.

Jelen, T.G., and Lockett, L.A. (2014). "Religion, Partisanship, and Attitudes toward Science Policy." *SAGE Open* 4(1).

Ji, F., and Zhong Y.G. (2010). "Is China Ready for Human Interference with the Climate?" *Pioneering with Science and Technology* 5: 30–35.

Jones, C.W. (2011). "CO_2 Capture from Dilute Gases as a Component of Modern Global Carbon Management." *Annu. Rev. Chem. Biomol. Eng.* 2: 31–52.

Jones, C., Lowe, J., Liddicoat, S., and Betts, R. (2009). "Committed Ecosystem Change due to Climate Change." *Nature Geoscience* 2: 484–487.

Jones, R., Rigg, C., and Lee, L. (2010). "Haida Marine Planning: First Nations as a Partner in Marine Conservation." *Ecology and Society* 15(1): 12.

Kahan, D., Silva, C., Tarantola, T., Jenkins-Smith, H., and Braman, D. (2015). "Geoengineering and Climate Change Polarization: Testing a Two-Channel Model of Science Communication." *Annals of American Academy of Political & Social Science* 658(1): 192–222.

Kahn, H., and Wiener, A.J. (1967). "The Next Thirty-Three Years: A Framework for Speculation." *Daedalus* 96(3): 705–732.

Karinen, R., and Guston, D.H. (2010). "Toward Anticipatory Governance: The Experience with Nanotechnology." In Kaiser, M., Kurath, M., Maasen, S., and Rehmann-Sutter, C. (eds.) *Governing Future Technologies: Nanotechnology and the Rise of an Assessment Regime*. Dordrecht, the Netherlands: Springer, pp. 217–232.

Keith, D.W. (2000). "Geoengineering the Climate: History and Prospect." *Annual Review of Energy and the Environment* 25: 245–284.

Keith, D.W. (2009a). "Engineering the Planet." In Schneider, S. and Mastrandrea, M. (eds.). *Climate Change Science and Policy.* Washington, DC: Island Press. pp. 494–592.

Keith, D.W. (2009b). "Why Capture CO_2 from the Atmosphere?" *Science* 325(5948): 1654–1655.

Keith, D.W. (2010). "Photophoretic Levitation of Engineered Aerosols for Geoengineering." *Proceedings of the National Academy of Sciences* 107(38): 16428–16431.

Keith, D.W. (2013). *A Case for Climate Engineering.* Boston, MA: MIT Press.

Keith, D.W., Duren, R., and MacMartin, D.G. (2014). "Field Experiments on Solar Geoengineering: Report of a Workshop Exploring a Representative Research Portfolio." *Philosophical Transactions of the Royal Society of London A: Mathematical, Physical and Engineering Sciences* 372(2031).

Keith, D.W., and MacMartin, D.G. (2015). "A Temperate, Moderate and Responsive Scenario for Solar Geoengineering." *Nature Climate Change* 5: 201–206. doi:10.1038/ nclimate2493.

Keith, D.W., Minh Ha, H.D., and Stolaroff, J.K. (2006). "Climate Strategy with CO_2 Capture from the Air." *Climatic Change* 74: 17–45.

Keith, D.W., Parson, E., and Morgan, M.G. (2010). "Research on Global Sun Block Needed Now." *Nature* 463(28): 426–427.

Keller, E.F. (1985). *Reflections on Gender and Science.* New Haven, CT: Yale University Press.

Keller, K., Bolker, B.M., and Bradford, D.F. (2004). "Uncertain Climate Thresholds and Optimal Economic Growth." *Journal of Environmental Economics and Management* 48(1): 723–741.

Keller, K., Hall, M., Kim, S.R., Bradford, D.F., et al. (2005). "Avoiding Dangerous Anthropogenic Interference with the Climate System." *Climatic Change* 73(3): 227–238.

Keller, K., and McInerney, D. (2008). "The Dynamics of Learning About a Climate Threshold." *Climate Dynamics* 30: 321–332.

Kelly, E.P. (2012). "An Archaeological Interpretation of Irish Iron Age Bog Bodies". In Ralph, S. (ed.). *The Archaeology of Violence: Interdisciplinary Approaches.* New York: State University of New York Press, pp. 232–240.

Kiehl, J. (2006). "Geoengineering Climate Change: Treating the Symptom over the Cause?" *Climatic Change* 77(3): 227–228.

Kintisch, E. (2010a). "Exclusive Excerpt: Hack the Planet." *WIRED.* March 24. Available at: http://www.wired.com/2010/03/hack-the-planet-excerpt/

Kintisch, E. (2010b). "EARTH: Emergency Procedures Safety Card." Available at: http://hacktheplanetbook.com/safetycard

Kintisch, E. (2010c). "Proposed Biodiversity Pact Bars 'Climate-Related Geoengineering.'" *Science Insider.* October 26. Available at: http://news.sciencemag. org/scienceinsider/2010/10/proposed-biodiversity-pact-bars-.html

Kirk-Davidoff, D.B., Hintsa, E.J., Anderson, J.G., and Keith, D.W. (1999). "The Effect of Climate Change on Ozone Depletion Through Changes in Stratospheric Water Vapor." *Nature* 402(6760): 399–401.

Kössler, G.P. (2012). *Geo-Engineering: Gibt es wirklich einen Plan(eten) B?* Berlin: Heinrich Böll Foundation. Available at: http://www.boell.de/de/content/geo-engineering-gibt-es-wirklich-einen-planeten-b

Kravitz, B., MacMartin, D.G., Robock, A., Rasch, P.J., Ricke, K.L., Cole, J.N.S., Curry, C.L., Irvine, P.J., Ji, D., Keith, D.W., Kristjánsson, J.E., Moore, J.C., Muri, H., Singh, B., Tilmes, S., Watanabe, S., Yang, S., and Yoon, J.-H. (2014). "A Multi-model Assessment of Regional Climate Disparities caused by Solar Geoengineering." *Environmental Research Letters* 9: 074013.

Kravitz, B., Forster, P.M., Jones, A., Robock, A., Alterskjær, K., Boucher, O., Jenkins, A.K.L., Korhonen, H., Kristjánsson, J.E., Muri, H., Niemeier, U., Partanen, A.-I., Rasch, P.J., Wang, H., and Watanabe, S. (2013). "Sea Spray Geoengineering Experiments in the Geoengineering Model Intercomparison Project (GeoMIP): Experimental Design and Preliminary Results." *Journal of Geophysical Research* 118(19): 11175–11186. doi:10.1002/jgrd.50856.

Kravitz, B., MacMartin, D.G., and Caldeira, K. (2012). "Geoengineering: Whiter Skies?" *Geophysical Research Letters* 39: L11801. doi: 10.1029/2012GL051652.

Kravitz, B., Robock, A., Boucher, O., Schmidt, H., Taylor, K.E., Stenchikov, G., and Schulz, M. (2011). "The Geoengineering Model Intercomparison Project (GeoMIP)." *Atmospheric Science Letters* 12(2): 162–167.

Kravitz, B., Robock, A., Forster, P.M., Haywood, J.M., Lawrence, M.G., and Schmidt, H. (2013b). "An Overview of the Geoengineering Model Intercomparison Project (GeoMIP)." *Journal of Geophysical Research* 118: 13103–13107. doi:10.1002/2013JD020569.

Kravitz, B., Robock, A. Tilmes, S., Boucher, O., English, J.M., Irvine, P.J., Jones, A., Lawrence, M.G., MacCracken, M., Muri, H., Moore, J.C., Niemeier, U., Phipps, S.J., Sillmann, J., Storelvmo, T., Wang, H., and Watanabe, S. (2015). "The Geoengineering Model Intercomparison Project Phase 6 (GeoMIP6): Simulation design and preliminary results." *Geoscientific Model Development* 8: 3379–3392. doi: 10.5194/gmd-8-3379-2015.

Kruger, T., Rayner, S., Redgwell, C., Savulescu, J., and Pidgeon, N. (2010). "Memorandum Submitted by T. Kruger et al." Available at: http://www.geoengineering.ox.ac.uk/oxford-principles/history/

Lackner, K. (2009). "Capture of Carbon Dioxide from Ambient Air." *European Physics Journal Special Topics* 176: 93–106.

Lackner, K.S., Brennan, S., Matter, J.M., Park, A.H., Wright, A., and van der Zwaan, B. (2012). "The Urgency of the Development of CO_2 Capture from Ambient Air." *Proceedings of the National Academy of Sciences* 109(33): 13156–13162.

Lackner, K.S., Grimes, P., and Ziock, H.J. (1999). "Carbon Dioxide Extraction from Air: Is it an Option?" *LAUR-99-5113*, Los Alamos Natl. Lab., Los Alamos, NM. Available at: http://www.gibbsenergy.com/gibbs_energy/Downloads_files/Extraction%20from%20Air.pdf

Laclau, E. (2006). *On Populist Reason*. London: Verso.

Lane, L. (2014). "Toward a Conservative Policy on Climate Change." *The New Atlantis* Winter: 19–37.

Langmuir, I. (1948). "The Growth of Particles in Smokes and Clouds and the Production of Snow from Supercooled Clouds." *Proceedings of the American Philosophical Society* 92: 167–185.

Latham, J. (1990). "Control of Global Warming?" *Nature* 347: 339–340.

Lavoie, J. (2012). "Residents Split over Iron Project." *Victoria Times Colonist*. 27 October. Available at: http://www.timescolonist.com/news/residents-split-over-iron-project-1.1014

Lawrence, M.G. (2006). "The Geoengineering Dilemma: To Speak or Not to Speak." *Climatic Change* 77(3–4): 245–248.

Lefale, P.F. (2001). *Climate Change Negotiations; Observations from the Frontline*. LLM Dissertation Paper (No. 1), International Environmental Law (LLM 708), Faculty of Law, University of Auckland, Auckland, New Zealand (unpublished).

Leinen, M.S. (2011). "The Asilomar International Conference on Climate Intervention Technologies: Background and Overview." *Stanford Journal of Law, Science, and Policy* 4(1): 1–5.

Lempert, R.J., and Collins, M.T. (2007). "Managing the Risk of Uncertain Threshold Responses: Comparison of Robust, Optimum, and Precautionary Approaches." *Risk Analysis* 27(4): 1009–1026.

Lempert, R., Nakicenovic, N., Sarewitz, D., and Schlesinger, M. (2004). "Characterizing Climate-Change Uncertainties for Decisionmakers." *Climatic Change* 65: 1–9.

Lenton, T.M. (2011a). "Early Warning of Climate Tipping Points." *Nature Climate Change* 1: 201–209.

Lenton, T.M. (2011b). "Beyond 2°C: Redefining Dangerous Climate Change for Physical Systems." *Wiley Interdisciplinary Reviews: Climate Change* 2(3): 451–461.

Lenton, T.M. (2018). "Can Emergency Geoengineering Really Prevent Climate Tipping Points?" In Blackstock, J.J. and Low, S. (eds.). *Geoengineering Our Climate? Ethics, Politics and Governance*. London: Earthscan from Routledge, pp. 43–46 (this volume).

Lenton, T.M., Held, H., Kriegler, E., Hall, J.W., Lucht, W., Rahmstorf, S., and Schellnhuber, H.J. (2008). "Tipping Elements in the Earth's Climate System." *Proceedings of the National Academy of Sciences* 105(6): 1786–1793.

Lenton, T.M. et al. (2009). "Using GENIE to Study a Tipping Point in the Climate System." *Philosophical Transactions of the Royal Society A* 367(1890): 871–884.

Lenton, T.M., and Vaughan, N. (2009). "The Radiative Forcing Potential of Different Climate Geoengineering Options." *Atmospheric Chemistry and Physics* 9: 2559–2608.

Leslie, E. (1842). "The Rain King, or, A Glance at the Next Century." *Godey's Lady's Book* 25: 7–11.

Lin, A. (2009). "Geoengineering Governance." *Issues in Legal Scholarship* 8(1): article 2.

Linner, B.O., and Wibeck, V. (2015). "Dual High Stake Emerging Technologies: A Review of the Climate Engineering Research Literature." *WIREs Climate Change* 6: 255–268. doi: 10.1002/wcc.333

Livina, V.N., and Lenton, T.M. (2007). "A Modified Method for Detecting Incipient Bifurcations in a Dynamical System". *Geophysical Research Letters* 34: L03712.

London Convention and London Protocol. (2010). Thirty-second Consultative Meeting of Contracting Parties to the London Convention and Eighth Meeting of Contracting Parties to the London Protocol, Oct. 11–15, Resolution LC-LP.2 on the Assessment Framework for Scientific Research Involving Ocean Fertilization. IMO Doc. LC 32/15/ Annex 5.

London Convention and London Protocol. (2013). Thirty-fifth Consultative Meeting of Contracting Parties to the London Convention and Eighth Meeting of Contracting Parties to the London Protocol, Oct. 14–18, Resolution LP.4(8) on the Amendment to the London Protocol to Regulate the Placement of Matter for Ocean Fertilization and Other Marine Geoengineering Activities. IMO Doc. LC 35/15/Annex 4.

Long, J.C.S., Hamburg, S.P., and Shepherd, J. (2012). "Climate: More Ways to Govern Geoengineering." *Nature* 486(7403): 323.

Long, J.C.S., and Winickoff, D. (2010). "Governing Geoengineering Research: Principles and Process." *The Solutions Journal* 1(5): 60–62.

Lord, R., Goldberg, S., Rajamani, L., and Brunnee, J. (eds.) (2013). *Climate Change Liability: Transnational Law and Practice*. New York: Cambridge University Press.

Low, S. (2017a). "The Futures of Climate Engineering." *Earth's Future* 5: 67–71.

Low, S. (2017b). "Engineering Imaginaries: Anticipatory Foresight for Solar Radiation Management Governance." *Science of the Total Environment* 580: 90–104.

Lukacs, M. (2012). "World's Biggest Geoengineering Experiment 'Violates' UN Rules." *The Guardian*. October 15. Available at: http://www.theguardian.com/ environment/2012/oct/15/pacific-iron-fertilisation-geoengineering

Lukacs, M., Goldenberg, S., and Vaughan, A. (2013). "Russia Urges UN Climate Report to Include Geoengineering." *The Guardian*. Available at: http://www.theguardian.com/environment/2013/sep/19/russia-un-climate-report-geoengineering

Lunt, D.J., Ridgwell, A., Valdes, P.J., and Seale, A. (2008). "'Sunshade World': A Fully Coupled GCM Evaluation of the Climatic Impacts of Geoengineering." *Geophysical Research Letters* 35: L12710.

MacCracken, M.C. (2006). "Geoengineering: Worthy of Cautious Evaluation?" *Climatic Change* 77(3–4): 235–243.

MacCracken, M., Barrett, S., Barry, R., Crutzen, P., Hamburg, S., Lampitt, R., Liverman, D., Lovejoy, T., McBean, G., Parson, E., Seidel, S., Shepherd, J., Somerville, R., and Wigley, T.M.L. (2010b). *The Asilomar Conference Recommendations on Principles for Research into Climate Engineering Techniques*. The Climate Institute. Available at: http://www.climate.org/PDF/AsilomarConferenceReport.pdf

MacCracken, M., Berg, P., Crutzen, P., Barrett, S., Barry, R., Hamburg, S., Lampitt, R., Liverman, D., Lovejoy, T., McBean, G., Parson, E., Seidel, S., Shepherd, J., Somerville, R., Wigley, T.M.L., et al. (2010a). "Conference Statement." The Climate Institute. Available at: http://www.climate.org/resources/climate-archives/conferences/asilomar/statement.html

MacKerron, G. (2014). "Costs and Economics of Geoengineering." *Climate Geoengineering Governance Working Paper Series*: 013.

Markus, T., and Ginzky, H. (2011). "Regulating Climate Engineering: Paradigmatic Aspects of the Regulation of Ocean Fertilization." *Carbon and Climate Law Review* 4: 477–490.

McCright, A.M., Dunlap, R.E., and Xiao, C. (2014). "The Impacts of Temperature Anomalies and Political Orientation on Perceived Winter Warming." *Nature Climate Change* 4: 1077–1081.

MacMartin, D.G., Keith, D.W., Kravitz, B., and Caldeira, K. (2013). "Management of Trade-offs in Geoengineering through Optimal Choice of Non-uniform Radiative Forcing." *Nature Climate Change* 3: 365–368. doi: 10.1038/nclimate1722.

MacMynowski, D.G., Shin, H.-J., and Caldeira, K. (2011). "The Frequency Response of Temperature and Precipitation in a Climate Model." *Geophysical Research Letters* 38(L16711). doi: 10.1029/2011GL048623.

Macnaghten, P., and B. Szerszynski. (2013). "Living the Global Social Experiment: An Analysis of Public Discourse on Solar Radiation Management and Its Implications for Governance." *Global Environmental Change* 23: 465–474.

Maibach, E., Roser, H., Renouf, C., and Leiserowitz, A. (2009). *Global Warming's Six Americas 2009: An Audience Segmentation Analysis*. Available at: www.climatechange.gmu.edu

Markusson, N., Ginn, F., Ghaleigh, N.S., and Scott, V. (2013). "In Case of Emergency Press Here: Framing Geoengineering as a Response to Dangerous Climate Change." *WIREs Climate Change* 5: 281–290.

Marshall, M. (2012). "Controversial Geoengineering Field Test Cancelled." *New Scientist*. May 22. Available at: http://www.newscientist.com/article/dn21840-controversial-geoengineering-field-test-cancelled.html#.VVuQ1flVhBc

Martineau, J. (1999). "Otter Skins, Clearcuts, and Ecotourists: Re-resourcing Haida Gwaii." In Miller, M., Auyong, J., and Hadley, N. (eds.). *Proceedings of the 1999 International Symposium on Coastal and Marine Tourism: Balancing Tourism and Conservation*. Vancouver, BC: Oceans Blue Foundation, pp. 237–249.

Martínez-Alier, J. (2002). *The Environmentalism of the Poor: A Study of Ecological Conflicts and Valuation*. Cheltenham, UK: Edward Elgar.

Matthews, H.D., and Caldeira, K. (2007). "Transient Climate-Carbon Simulations of Planetary Geoengineering." *Proceedings of the National Academy of Sciences* 104(24): 9949–9954.

McCain, L. (2002). "Informing Technology Policy Decisions: The US Human Genome Project's Ethical, Legal, and Social Implications Programs as a Critical Case." *Technology and Society* 24(1–2): 111–132.

McClellan, J., Keith, D., and Apt, J. (2010). "Cost Analysis of Stratospheric Albedo Modification Delivery Systems." *Environmental Research Letters* 7: 034019

McClellan, J., Sisco, J., Suarez, B., and Keogh, G. (2011). *Geoengineering Cost Analysis: Final Report.* Aurora Flight Services, prepared under Contract to the University of Calgary, report AR10-182

McCormick, M.P., Thomason, L.W., and Trepet, C.R. (1995). "Atmospheric Effects of the Mt Pinatubo Eruption." *Nature* 373: 399–404.

McGlashan, N., Shah, N., Caldecott, B., and Workman, M. (2012). "High-level Techno-economic Assessment of Negative Emissions Technologies." *Process Safety and Environmental Protection* 90(6): 501–510.

McGlashan, N., Shah, N., and Workman, M. (2010). *The Potential for the Deployment of Negative Emissions Technologies in the UK.* Work stream 2, Report 18 of the AVOID programme (AV/WS2/D1/R18). Available at www.avoid.uk.net

McKibben, B. (2013). "Bill McKibben on Geoengineering—I'm Annoyed." Video message prepared for the Forum for Climate Engineering Assessment. Available at http://dcgeoconsortium.org/2013/12/19/bill-mckibben-on-geoengineering-im-annoyed/

McKnight, Z. (2013). "Why Was Iron Dumping a Surprise?" September 3. Available at: http://www.vancouversun.com/technology/iron+dumping+surprise/8865130/story.html

McLaren, D.P. (2012a). "A Comparative Global Assessment of Potential Negative Emissions Technologies." *Process Safety and Environmental Protection* 90(6): 489–500.

McLaren, D.P. (2012b). "Procedural Justice in Carbon Capture and Storage." *Energy & Environment* 23(2–3): 345–365.

Marx, L. (1983). "Are Science and Society going in the Same Direction?" *Science, Technology and Human Values* 8(4): 6–9.

Meehl, G.A., Stocker, T.F., Collins, W.D., Friedlingstein, P., Gaye, A.T., Gregory, J.M., Kitoh, A., Knutti, R., Murphy, J.M., Noda, A., Raper, S.C.B., Watterson, I.G., Weaver, A.J., and Zhao, Z.C. (2007). In Solomon, S., Qin, D., Manning, M., Chen, Z., Marquis, M., Averyt, K.B., Tignor, M., and Miller, H.L. (eds.). *IPCC Climate Change 2007: The Physical Science Basis: Contribution of Working Group I to the Fourth Assessment Report of the Intergovernmental Panel on Climate Change.* Cambridge: Cambridge University Press, pp. 747–846.

Meleshko, V.P., Kattsov, V.M., and Karol, I.L. (2010). "Is Aerosol Scattering in the Stratosphere a Safety Technology Preventing Global Warming?" *Russian Meteorology and Hydrology* 35(7): 433–440.

Mendler de Suarez, J., Suarez, P., Bachofen, C., Fortugno, N., Goentzel, J., Gonçalves, P., Grist, N., Macklin, C., Pfeifer, K., Schweizer, S., Van Aalst, M., and Virji, H. (2012). "Games for a New Climate: Experiencing the Complexity of Future Risks." Pardee Center Task Force Report. Boston: The Frederick S. Pardee Center for the Study of the Longer-Range Future, Boston University.

Mercado, L.M., Bellouin, N., Sitch, S., Boucher, O., Huntingford, C., Wild, M. and Cox, P.M. (2009). "Impact of Changes in Diffuse Radiation on the Global Land Carbon Sink." *Nature* 458: 1014–1017.

Merchant, B. (2014). "The Climate Scientist who Pioneered Geoengineering Fears It's About to Blow Up." *Motherboard.* 25 August. Available at: http://motherboard.vice.com/read/ken-caldeira-climate-geoengineering

Mercer, A., Keith, D. and Sharp, J. (2011). "Public Understanding of Solar Radiation Management." *Environmental Research Letters* 6: 1–9.

Merchant, C. (1980). *The Death of Nature: Women, Ecology, and the Scientific Revolution.* San Francisco: Harper and Row.

Michaelson, J. (1998). "Geoengineering: A Climate Change Manhattan Project." *Stanford Environmental Law Journal* 17: 73–140.

Millard-Ball, A. (2012). "The Tuvalu Syndrome." *Climatic Change* 110(3–4): 1047–1066.

Mimura, N., Nurse, L., McLean, R.F., Agard, J., Briguglio, L., Lefale, P., Payet, R., and Sem, G. (2007). *Small Islands.* In Parry, M.L., Canziani, O.F., Palutikof, J.P., van der Linden, P.J., and Hanson, C.E. (eds.). Climate Change 2007: Impacts, Adaptation and Vulnerability. Contribution of Working Group II to the Fourth Assessment Report of the Intergovernmental Panel on Climate Change. Cambridge: Cambridge University Press. pp. 687–716. Available at: http://www.ipcc.ch/pdf/assessment-report/ar4/wg2/ar4-wg2-chapter16.pdf

Moan, J., Porojnicu, A.C., Dahlback, A., and Setlowm, R.B. (2008). "Addressing the Health Benefits and Risks, Involving Vitamin D or Skin Cancer, of Increased Sun Exposure." *Proceedings of the National Academy of Sciences* 105(2): 668–673.

Mooney, C. (2009). "Copenhagen: Geoengineering's Big Break?" *Mother Jones.* Available at: http://www.motherjones.com/environment/2009/12/copenhagen-geoengineerings-big-break

Mooney, C. (2010). *Do Scientists Understand the Public?* Cambridge, MA: American Academy of Arts and Sciences.

Moore, J.C., Chen, Y., Cui, X.F., Yuan, W.P., Dong, W.J., Gao, Y., and Shi, P.J. (2016). "Will China Be the First to Initiate Climate Engineering?" *Earth's Future* 4(12): 588–595.

Moreno-Cruz, J. (2015). "Mitigation and the Geoengineering Threat." *GT SOE Working Paper.* Available at: http://works.bepress.com/morenocruz/3/

Moreno-Cruz, J., and Keith, D.W. (2012). "Climate Policy under Uncertainty: A Case for Solar Geoengineering." *Climatic Change* 121(3): 431–444. doi: 10.1007/s10584-012-0487-4.

Moreno-Cruz, J., Ricke, K.L., and Keith, D.W. (2011). "A Simple Model to Account for the Regional Inequalities in the Effectiveness of Solar Radiation Management." *Climatic Change* 110: 649–668.

Morgan, M.G., Kandlikar, M., Risbey, J., and Dowlatabadi, H. (1999). "Why Conventional Tools for Policy Analysis Are Often Inadequate for Problems of Global Change. *Climatic Change* 41(3): 271–281.

Morgan, M.G., Nordhaus, R.R., and Gottlieb, P. (2013). "Needed: Research Guidelines for Solar Radiation Management." *Issues in Science and Technology* Spring: 37–44.

Morgan, M.G., and Ricke, K.L. (2010). "Cooling the Earth Through Solar Radiation Management: The Need for Research and an Approach to its Governance." Technical Report. *International Risk Governance Council.*

Moriyama, R., Sugiyama, M., Kurosawa, A., Masuda, K., Tsuzuki, K., and Ishimoto, Y. (2016). "The Cost of Stratospheric Climate Engineering Revisited." *Mitigation and Adaptation Strategies for Global Change* 22(8): 1207–1228.

Morrow, D.R., Kopp, R.E., and Oppenheimer, M. (2009). "Toward Ethical Norms and Institutions for Climate Engineering Research." *Environmental Research Letters*, 4, 045106.

Morrow, D.R., Kopp, R.E., and Oppenheimer, M. (2013). "Political Legitimacy in Decisions about Experiments in Solar Radiation Management." In Burns, W.C.G. and Strauss, A. (eds.) *Climate Change Geoengineering: Philosophical Perspectives, Legal Issues, and Governance Frameworks.* Cambridge: Cambridge University Press.

MosNews. (2005). "Russian Scientist Suggests Burning Sulfur in Stratosphere to Fight Global Warming." MosNews. November 30. Available at: http://www.prometeus. nsc.ru/eng/science/scidig/05/dec.ssi#3

Muchembled, R. (1985). *Popular Culture and Elite Culture in France, 1400–1750.* Cochrane L. (trans.). Baton Rouge, LA: Louisiana State University Press.

Murphy, D.M. (2009). "Effect of Stratospheric Aerosols on Direct Sunlight and Implications for Concentrating Solar Power." *Environment Science Technology* 48(8): 2784–2786. doi: 10.1021/es802206b.

Müller, B. (1999). *Justice in Global Warming Negotiations: How to Obtain a Procedurally Fair Compromise.* Oxford: Oxford Institute for Energy Studies.

Myers, R. (2011). "The Public Values Failures of Climate Science in the US." *Minerva* 49(1): 47–70.

Naik, V., Wuebbles, D.J., DeLucia, E.H., and Foley, J.A. (2003). "Influence of Geoengineered Climate on the Terrestrial Biosphere." *Environmental Management* 32(3): 373–381.

Najam, A. (2005). "Why Environmental Politics Looks Different from the South." In P. Dauvergne (ed). *Handbook of Global Environmental Politics.* Cheltenham, UK: Edward Elgar.

Narisma, G.T., Foley, J.A., Licker, R., and Ramankutty, N. (2007). "Abrupt Changes in Rainfall During the Twentieth Century." *Geophysics Research Letters* 24(06), L06710.

National Academies of Sciences. (2009). "Advancing the Science of Climate Change." Available at: http://nas-sites.org/americasclimatechoices/sample-page/panel-reports/ 87-2/

National Commission for the Protection of Human Subjects of Biomedical and Behavioral Research. (1979). *The Belmont Report: Ethical principles and guidelines for the protection of human subjects of research.* Washington, DC: Dept. of Health, Education, and Welfare.

National Natural Science Foundation of China. (2012). "NSFC Project Guidelines (Earth Sciences Section)." *Advances in Earth Science* 27(1): 1–13.

National Oceanic and Atmospheric Administration (NOAA). (2010). The NOAA Annual Greenhouse Gas Index (AAGI). Available at: http://www.esrl.noaa.gov/ gmd/aggi/

National Research Council (NRC). (2015a). *Climate Intervention: Carbon Dioxide Removal and Reliable Sequestration.* Washington, DC: National Academies Press. https://doi. org/10.17226/18805

National Research Council (NRC). (2015b). *Climate Intervention: Reflecting Sunlight to Cool Earth.* Washington, DC: National Academies Press. https://doi.org/10.17226/18988

Natural Environment Research Council. (2010). *Experiment Earth? Report on a Public Dialogue on Geoengineering.* Available at: nerc.ac.uk/about/consult/geoengineering.asp

Nerlich, B., and R. Jaspal. (2012). "Metaphors We Die By? Geoengineering, Metaphors and the Argument from Catastrophe." *Metaphor and Symbol* 27(2): 131–147.

Niang, I., Ruppel, O.C., Abdrabo, M.A., Essel, A., Lennard, C., Padgham, J., and Urquhart, P. (2014). "Africa." In Barros, V.R., Field, C.B., Dokken, D.J., Mastrandrea,

M.D., Mach, K.J., Bilir, T.E., Chatterjee, M., Ebi, K.L., Estrada, Y.O., Genova, R.C., Girma, B., Kissel, E.S., Levy, A.N., MacCracken, S., Mastrandrea, P.R., and White, L.L. (eds.) Climate Change 2014: Impacts, Adaptation, and Vulnerability. Part B: Regional Aspects. Contribution of Working Group II to the Fifth Assessment Report of the Intergovernmental Panel on Climate Change. Cambridge and New York: Cambridge University Press, pp. 1199–1265.

Nicholson, S., Jinnah, S., and Gillespie, A. (2017). "Solar Radiation Management: A Proposal for Immediate Polycentric Governance." *Climate Policy* 18(3): 322–334.

Nisbet, M. (2009). "Communicating Climate Change: Why Frames Matter for Public Engagement." *Environment* 51(2): 12–23.

Noble, D.F. (1999). *The Religion of Technology: The Divinity of Man and the Spirit of Invention.* London: Penguin.

Nordhaus, W.D. (1979). *The Efficient Use of Energy Resources.* New Haven, CT: Yale University Press.

Novim (2009). Blackstock, J.J., Battisti, D.S., Caldeira, K., Eardley, D.M., Katz, J.I., Keith, D.W., Patrinos, A.A.N., Schrag, D.P., Socolow, R.H., and Koonin, S.E. *Climate Engineering Responses to Climate Emergencies.* July. Available at: http://arxiv.org/pdf/0907.5140

Nurse, L. McLean, R.F., Agard, J., Briguglio, L., Duvat, V., Pelesikoti, N., Tompkins, E., and Webb, A. (2014). *Small Islands.* In Barros, V.R., Field, C.B., Dokken, D.J., Mastrandrea, D.J., Biler, T.E., Chatterjee, M., Ebi, K.L., Estrada, Y.L., Genova, R.C., Girma, B., Kissel, E.S., Levy, A.N., MacCracken, S., Mastrandrea, P.N., and White, L.L. Climate Change 2014: Impacts, Adaptation and Vulnerability. Contribution of Working Group II to the Fifth Assessment Report of the Intergovernmental Panel on Climate Change. Cambridge: Cambridge University Press. pp. 1613–1654. Available at: http://www.ipcc.ch/pdf/assessment-report/ar5/wg2/WGIIAR5-Chap29_FINAL.pdf

Oman, L., Robock, A., Stenchikov, G.L., and Thordarson, T. (2006). "High-Latitude Eruptions Cast Shadow over the African Monsoon and the Flow of the Nile." *Geophysical Research Letters* 33: L18711. doi: 10.1029/2006GL027665.

ORNL Review. (2002). Director Alvin Weinberg. ORNL Review 25/3&4.

Oschlies, A., Held, H., Keller, D., Keller, K., Mengis, N., Quaas, M., Rickels, W., and Schmidt, H. (2016). "Indicators and Metrics for the Assessment of Climate Engineering." *Earth's Future* 5(1): 49–58 doi: 10.1002/2016EF000449.

Parker, A. (2014). "Governing Solar Geoengineering Research as It Leaves the Laboratory." *Philosophical Transactions of the Royal Society A* 372: 20140173.

Parfit, D. (1997). "Equality and Priority." *Ratio* 10: 202–221.

Parkhill, K., and Pidgeon, N. (2011). "Public Engagement on Geoengineering Research: Preliminary Report on the SPICE Deliberative Workshops." *Working Paper. Understanding Risk Research Group.* Cardiff University. pp. 1–29.

Parson, E.A. and Ernst L.N. (2013). "International Governance of Climate Engineering." *Theoretical Inquiries in Law* 14: 307–337.

Parson, E.A. and Keith, D.W. (2013). "End the Deadlock on Governance of Geoengineering Research." *Science* 339: 1278–1279.

Parsons, T., and Whitney, F. (2012). "Did Volcanic Ash from Mt. Kasatochi in 2008 Contribute to a Phenomenal Increase in Fraser River Sockeye Salmon (Oncorhynchus nerka) in 2010?" *Fisheries Oceanography* 21(5): 374–377.

Pereira, J.C. (2016). "Geoengineering, Scientific Community, and Policymakers: A New Proposal for the Categorization of Responses to Anthropogenic Climate Change." *SAGE Open.* doi: org/10.1177/2158244016628591.

Peters, T. (2007). "Are We Playing God with Nanoenhancement?" In Allhoff, F. (ed.). *Nanoethics: The Ethical and Social Implications of Nanotechnology.* Hoboken, NJ: John Wiley & Sons.

Petersen, A. (2018). "The Emergence of the Geoengineering Debate Within the IPCC." In Blackstock, J.J. and Low, S. (eds.). *Geoengineering Our Climate? Ethics, Politics and Governance.* London: Earthscan from Routledge, pp. 121–124 (this volume).

Pew Research Center. (2012). *The Global Religious Landscape.* Washington, DC: Pew Research Religion & Public Life Project. Available at: http://www.pewforum.org/files/2014/01/global-religion-full.pdf

Pidgeon, N., Corner, A., Parkhill, K., Spence, A., and Butler, C. (2012). "Exploring Early Public Responses to Geoengineering." *Philosophical Transactions of the Royal Society A* 370: 4176–4196.

Pidgeon, N., Parkhill, K., Corner, A., and Vaughan, N. (2013). "Deliberating Stratospheric Aerosols for Climate Geoengineering and the SPICE Project." *Nature Climate Change* 3(5): 451–457.

Pielke Jr., R.A. (1995). "Usable Information for Policy: An Appraisal of the U.S. Global Change Research Program." *Policy Science* 38: 39–77.

Pielke Jr., R.A., Klein, R., and Sarewitz, D. (2000). "Turning the Big Knob: An Evaluation of the Use of Energy Policy to Modulate Future Climate Impacts." *Energy and Environment* 11: 255–276.

Pitari, G., Aquila, V., Kravitz, B., Robock, A., Watanabe, S., Cionni, I., De Luca, N., Di Genova, G., Mancini, E., and Tilmes, S. (2014). "Stratospheric Ozone Response to Sulfate Geoengineering: Results from the Geoengineering Model Intercomparison Project." *Journal of Geophysical Research: Atmospheres* 119: 2629–2653.

Porter, K., and Hulme, M. (2013). "The Emergence of the Geoengineering Debate in the UK Print Media: a Frame Analysis." *Geographical Journal* 179: 342–355.

Posner, E.A., and Weisbach, D. (2010). *Climate Change Justice.* Princeton, NJ: Princeton University Press.

"Press Review: Iron Fertilization in Canada/Haida Gwaii (7th update, 05.09.'13)." (2013). Website. *Climate Engineering.* Kiel Earth Institute and Heidelberg Center for the Environment. Available at: http://www.climate-engineering.eu/single/items/press-review-iron-fertilization-in-canada.html

Preston, C.J. (2011). "Re-Thinking the Unthinkable: Environmental Ethics and the Presumptive Argument against Geoengineering." *Environmental Values* 20: 457–479.

Preston C. (ed.) (2012). *Engineering the Climate: The Ethics of Solar Radiation Management.* Lanham, MD: Lexington Press.

Preston C. (2012). "Solar Radiation Management and Vulnerable Populations: the Moral Deficit and Its Prospects." In: Preston C. (ed.). *Engineering the Climate: The Ethics of Solar Radiation Management.* Lanham, MD: Lexington Press, pp. 77–94.

Rahmstorf, S. (1995). "Bifurcations of the Atlantic Thermohaline Circulation in Response to Changes in the Hydrological Cycle." *Nature* 378: 145–149.

Ramanathan, V. et al. (2005). "Atmospheric Brown Clouds: Impacts on South Asian Climate and Hydrological Cycle." *Proceedings of the National Academy of Sciences USA* 102(15): 5326–5333.

Rasch, P.J., Tilmes, S., Turco, R.P., Robock, A., et al. (2008). "An Overview of Geoengineering of Climate Using Stratospheric Sulphate Aerosols." *Philosophical Transactions of the Royal Society A – Mathematical Physical and Engineering Sciences* 366(1882): 4007–4037.

Rau, G. (2014). "Enhancing the Ocean's Role in CO_2 Mitigation." In Freeman, B. (ed.). *Handbook of Global Environmental Change.* New York: Springer.

Raven, J., Caldeira, K., Elderfield, H., Hoegh-Guldberg, O., et al. (2005). *Ocean Acidification Due to Increasing Atmospheric Carbon Dioxide.* London: Royal Society.

Ravetz, J., and Funtowicz, S. (1993). "Science for the Post-Normal Age." *Futures* 25(7): 739–755.

Rawls, J. (1971). *A Theory of Justice.* Cambridge, MA: Harvard University Press.

Ray, J. (2003). *Worlds Apart: Religion in Canada, Britain, U.S.* Washington, DC: Gallup. Available at: http://www.gallup.com/poll/9016/worlds-apart-religion-canada-britain-us.aspx

Rayfuse, R., Lawrence, M.G., and Gjerde, K.M. (2008). "Ocean Fertilization and Climate Change: The Need to Regulate Emerging High Seas Uses." *International Journal of Marine and Coastal Law* 23(2): 297–326.

Rayner, S. (2004). "The Novelty Trap: Why Does Institutional Learning about New Technologies Seem So Difficult?" *Industry and Higher Education* 18(6) December: 349–355.

Rayner, S., Heyward, C., Kruger, T., Pidgeon, N., Redgwell, C., and Savulescu, J. (2013). "The Oxford Principles." *Climatic Change* 121(3): 499–512.

Rayner, S, Redgwell, C., Savulescu, J., Pidgeon, N., and Kruger, T. (2009). "Memorandum on Draft Principles for the Conduct of Geoengineering Research." Available at: http://www.geoengineering.ox.ac.uk/oxford-principles/history/

Revkin, A. (2014). "A Darker View of the Age of Us – the Anthropocene." *New York Times Dot Earth Blog.* June 18. Available at: http://dotearth.blogs.nytimes.com/2014/06/18/a-darker-view-of-the-age-of-us-the-anthropocene/?_r=0

Reynolds, J.L. (2014a). "Climate Engineering Field Research: The Favorable Setting of International Environmental Law." *Washington and Lee Journal of Energy, Climate, and the Environment* 5(2): 417–486.

Reynolds, J. (2014b). "The International Regulation of Climate Engineering: Lessons from Nuclear Power." *Journal of Environmental Law* 26(2): 269–289.

Reynolds, J.L. (2015). "An Economic Analysis of Liability and Compensation for Harm from Large-Scale Field Research in Solar Climate Engineering." *Climate Law* 5(2–4): 182–209.

Reynolds, J.L. (2018). "The International Legal Framework for Climate Engineering." In Blackstock, J.J. and Low, S. (eds.). *Geoengineering Our Climate? Ethics, Politics and Governance.* London: Earthscan from Routledge, pp. 124–136 (this volume).

Reynolds, J.L., Contreras, J.L., and Sarnoff, J.D. (2017). "Solar Climate Engineering and Intellectual Property: Toward a Research Commons." *Minnesota Journal of Law, Science & Technology* 18(1): 1–110.

Ricke, K.L., Moreno-Cruz, J., and Caldeira, K. (2013). "Strategic Incentives for Climate Coalitions to Exclude Broad Participation." *Environmental Research Letters* 8(1): 014021.

Ricke, K.L., Morgan, M.G., and Allen, M.R. (2010). "Regional Climate Response to Solar Radiation Management." *Nature Geosciences* 3: 537–541.

Rickels, W., Klepper, G., Dovern, J., Betz, G., Brachatzek, N., Cacean, S., Güssow, K., Heintzenberg, J., Hiller, S., Hoose, C., Leisner, T., Oschlies, A., Platt, U., Proelß, A., Renn, O., Schäfer, S., and Zürn, M. (2011). *Large-Scale Intentional Interventions into the Climate System? Assessing the Climate Engineering Debate.* Kiel, Germany: Kiel Earth Institute. Available at: http://www.kiel-earth-institute.de/scoping-report-climate-engineering.html

Ridley, J., Gregory, J., Huybrechts, P., and Lowe, J. (2009). "Thresholds for Irreversible Decline of the Greenland Ice Sheet." *Climate Dynamics* 35(6): 1049–1057.

Robock, A. (2008). "20 Reasons Why Geoengineering May Be a Bad Idea." *Bulletin of the Atomic Science* 64(2): 14–18.

Robock, A. (2012). "Is Geoengineering Research Ethical?" *Peace and Security* 4: 226–229.

Robock, A., Bunzl, M., Kravitz, B., and Stenchikov, G. (2010). "A Test for Geoengineering?" *Science* 327: 530–531. doi: 10.1126/science.1186237.

Robock, A., and Kravitz, B. (2018). "Use of Models, Analogues and Field-tests for Geoengineering Research." In Blackstock, J.J. and Low, S. (eds.). *Geoengineering Our Climate? Ethics, Politics and Governance*. London: Earthscan from Routledge, pp. 95–99 (this volume).

Robock, A., MacMartin, D.G., Duren, R., and Christensen, M.W. (2013). "Studying Geoengineering with Natural and Anthropogenic Analogs." *Climatic Change*. doi:10.1007/s10584-013-0777-5.

Robock, A., Marquardt, A., Kravitz, B., and Stenchikov, G. (2009). "Benefits, Risks, and Costs of Stratospheric Geoengineering." *Geophysical Research Letters* 36: L19703.

Robock, A., Oman, L., and Stenchikov, G.L. (2008). "Regional Climate Responses to Geoengineering with Tropical SO_2 Injections." *Journal of Geophysical Research--Atmospheres*, 113(D16).

Rogelj, J., Meinshausen, M., and Knutti, R. (2012). "Global Warming under Old and New Scenarios Using IPCC Climate Sensitivity Range Estimates." *Nature Climate Change* 2: 248–253.

Royal Society (2009). Shepherd, J., Caldeira, K., Haigh, J., Keith, D., Launder, B., Mace, G., MacKerron, G., Pyle, H., Rayner, S., Redgwell, C., and Watson, A. *Geoengineering the Climate: Science, Governance and Uncertainty*. London: Royal Society.

Russell, L.M. (2012). "Offsetting Climate Change by Engineering Air Pollution to Brighten Clouds." *Bridge* 42(4): 10–15.

Russell, L.M., Sorooshian, A., Seinfeld, J.H., Albrecht, B.A., Nenes, A., Ahlm, L., Chen, Y.-C., Coggon, M., Craven, J.S., Flagan, R.C., Frossard, A.A., Jonsson, H., Jung, E., Lin, J.J., Metcalf, A.R., Modini, R., Mülmenstädt, J., Roberts, G., Shingler, T., Song, S., Wang, Z., and Wonaschütz, A. (2013). "Eastern Pacific Emitted Aerosol Cloud Experiment." *Bulletin of the American Meteorological Society* 94: 709–729.

Sarewitz, D. (2011). "Anticipatory Governance of Emerging Technologies." In Marchant, G., Allenby, B., and Herkert, J. (eds.). *The Growing Gap Between Emerging Technologies and Legal-Ethical Oversight: The Pacing Problem*. Dortrecht, the Netherlands: Springer. pp. 95–106.

Sarewitz, D., and Pielke Jr., R.A. (2008). "The Steps Not Yet Taken." In Kleinman, D., Cloud-Hansen, K., Matta, C., and Handelsman, J. (eds.). *Controversies in Science and Technology, Vol. 2, From Climate to Chromosomes*. New York: Mary Ann Liebert. pp. 329–351.

Scheffer, M. et al. (2009). "Early Warning Signals for Critical Transitions." *Nature* 461: 53–59.

Scheufele, D.A., Corley, E.A., Shih, T.J., Dalrymple, K.E., and Ho, S.S. (2009). "Religious Beliefs and Public Attitudes Toward Nanotechnology in Europe and the United States." *Nature Nanotechnology* 4: 91–94.

Schiermeier, Q. (2009). "Ocean Iron Fertilization Experiment Draws Fire." *Nature*. January 9. Available at: http://www.nature.com/news/2009/090109/full/news.2009.13.html

Schlenker, W., and Lobell, D.B. (2010). "Robust Negative Impacts of Climate Change on African Agriculture." *Environmental Research Letters* 5(1): 014010.

Schmidt, H., Alterskjaer, K., Bou Karam, D., Boucher, O., Jones, A., Kristjansson, J.E., Niemeier, U., Schulz, M., Aaheim, A., Benduhn, F., Lawrence, M., and Timmreck, C. (2012). "Solar Irradiance Reduction to Counteract Radiative Forcing from a Quadrupling of CO_2: Climate Responses Simulated by Four Earth System Models." *Earth System Dynamics* 3: 63–78.

Scholte, S., Vasileiadou, E., and Petersen, A.C. (2013). "Opening Up the Societal Debate on Climate Engineering: How Newspaper Frames Are Changing." *Journal of Integrative Environmental Sciences* 10(1): 1–16.

Schot, J., and Rip, A. (1997). "The Past and Future of Constructive Technology Assessment." *Technological Forecasting and Social Change* 54: 251–268.

Sclove, R. (1995). *Democracy and Technology.* New York: Guilford Press.

Scott, D. (ed.) (2012). "Special Issue on the Ethics of Geoengineering: Investigating the Moral Challenges of Solar Radiation Management." *Ethics, Policy and Environment* 15(2): 133–135.

Scott, K. (2005). "Day after Tomorrow: Ocean CO_2 Sequestration and the Future of Climate Change." *Georgetown International Environmental Law Review* XVIII: 57–108.

Seager, J. (2009). "Death by Degrees: Taking a Feminist Hard Look at the 2 Degrees Climate Policy." *Kvinder, Kon & Forskning -- Women, Gender & Research* 18(3–4): 11–22.

Secretariat of the Convention on Biological Diversity. (2009). *Scientific Synthesis of the Impacts of Ocean Fertilization on Marine Biodiversity.* Montreal, Canada. Technical Series No. 45.

Secretariat of the Convention on Biological Diversity. (2012). *Geoengineering in Relation to the Convention on Biological Diversity: Technical and Regulatory Matters.* Montreal, Canada. Technical Series No. 66.

Secretariat of the International Law Commission. (1995). *Survey on Liability Regimes Relevant to the Topic of International Liability for Injurious Consequences Arising Out of Acts Not Prohibited by International Law.* New York: United Nations.

Shackley, S. (2013). "EuTRACE – Studying GeoEngineering Options for the European Commission." *Environmental Governance.* Weblog. Available at: https://envirogov.wordpress.com/2013/09/24/eutrace-studying-geoengineering-options-for-the-european-commission/

Shapira, P., Youtie, J., and Porter, A. (2010). "The Emergence of Social Science Research in Nanotechnology." *Scientometrics* 85(2): 595–611.

Shukman, D. (2014). "Geo-engineering: Climate Fixes 'Could Harm Billions'." *BBC News.* November 26. Available at: http://www.bbc.com/news/science-environment-30197085

Sidgwick, H. (1907). *The Methods of Ethics, 7th ed.* London: Macmillan.

Sikka, T. (2012). "A Critical Discourse Analysis of Geoengineering Advocacy." *Critical Discourse Studies* 9: 163–175.

Silk, M. (2008). *One Nation, Divisible: How Regional Religious Differences Shape American Politics.* Lanham, MD: Rowman & Littlefield.

Silver, L.M. (2006). *Challenging Nature: The Clash of Science and Spirituality at the New Frontiers of Life.* New York: Ecco.

Smetacek, V. (2012). "Deep Carbon Export from a Southern Ocean Iron-fertilized Diatom Bloom." *Nature* 487: 313–319.

Smil, V. (2004). *Enriching the Earth: Fritz Haber, Carl Bosch, and the Transformation of World Food Production.* Cambridge, MA: MIT Press.

Smil, V. (2011). "Nitrogen Cycle and World Food Production." *World Agriculture* 2: 9–10.

Smith, P. et al. (2015). "Biophysical and Economic Limits to Negative CO_2 Emissions." *Nature Climate Change* 6(1): 42–50. doi: 10.1038/nclimate2870.

Soden, B.J., Wetherald, R.T., Stenchikov, G.L. and Robock, A. (2002). "Global Cooling After the Eruption of Mount Pinatubo: A Test of Climate Feeback by Water Vapor." *Science* 296: 727–730.

Solar Radiation Management Governance Initiative. *SRMGI Events.* Website. Available at: http://www.srmgi.org/events/

Solar Radiation Management Governance Initiative (SRMGI). (2013). *Governance of Research on Solar Geoengineering: African Perspectives.* Available at: http://www.aasciences.org/attachments/article/239/Governance-of-SRM-African-Perspectives.pdf

Solar Radiation Management Governance Initiative (SRMGI) (2016). *Solar Radiation Management: The Governance of Research.* Report. Available at: http://www.srmgi.org/files/2016/02/SRMGI.pdf

Steinbruner, J.D. (1997). "Biological Weapons: A Plague upon All Houses." *Foreign Policy* 109: 85–96.

Stilgoe, J. (2007). *Nanodialogues: Experiments in Public Engagement with Science.* London: Demos.

Stilgoe, J. (2015). *Experiment Earth: Responsible Innovation in Geoengineering.* London: Earthscan from Routledge.

Stilgoe, J., Owen, R., and Macnaghten, P. (2013a). "Developing a Framework for Responsible Innovation." *Research Policy* 42: 1568–1580.

Stilgoe, J., Watson, M., and Kuo, K. (2013b). "Public Engagement with Biotechnologies Offers Lessons for the Governance of Geoengineering Research and Beyond." *PLoS Biology* 11: e1001707.

Stirling, A. (2008). "'Opening Up' and 'Closing Down': Power, Participation, and Pluralism in the Social Appraisal of Technology." *Science, Technology and Human Values* 33: 262–294.

Stirling, A. (2014). "Emancipating Transformations: From Controlling 'the Transition' to Culturing Plural Radical Program." *Climate Geoengineering Governance Working Paper Series*: 012. Available at: http://steps-centre.org/publication/emancipating-transformations-controlling-transition-culturing-plural-radical-progress/

Stirling, A., and Mayer, S. (2001). "A Novel Approach to the Appraisal of Technological Risk: A Multicriteria Mapping Study of a Genetically Modified Crop." *Environment and Planning C: Government and Policy* 19: 529–555.

Stommel, H. (1961). "Thermohaline Convection with Two Stable Regimes of Flow." *Tellus* 13: 224–230.

Strauss, S.H., Tan, H., Boerjan, W. and Sedjo, R. (2009). "Strangled at Birth? Forest Biotech and the Convention on BIological Diversity." *Nature Biotechnology* 27(6): 519–527.

Strong, A.E. (1984). "Monitoring El Chichón Aerosol Distribution using NOAA-7 Satellite AVHRR Sea Surface Temperature Observations." *Geofisica International* 23: 129–141.

Strong, A., Chisholm, S., and Cullen, J. (2009). "Ocean Fertilization: Science, Policy, and Commerce." *Oceanography* 22(3): 236–261.

Suarez, P., Benn, J., and Macklin, C. (2011). "Putting Vulnerable People at the Center of Communication for Adaptation: The Case for Knowledge Sharing through Participatory Games and Video Tools." *World Resources Report 2011 – Expert Perspectives.*

Suarez, P., Blackstock, J.J., and Van Aalst, M. (2010). "Towards a People-centered Framework for Geoengineering Governance: A Humanitarian Perspective." *Geoengineering Quarterly* 1(1): 2–4.

Sugiyama, M., and Sugiyama, T. (2010). "Interpretation of CBD COP10 Decision on Geoengineering." *SERC Discussion Paper 10013*. Socio-economic Research Institute, Central Research Institute of Electric Power Industry.

Sullivan, P.R. (1995). "Murphy's Law and the Natural Ought." *Behavior and Philosophy* 24(1): 39–49.

Sutton, M.A., Oenema, O., Erisman, J.W., Leip, A., Van Grinsven, H., and Winivarter, W. (2011). "Too Much of a Good Thing." *Nature* 472: 159–161.

Svoboda, T. (2017). *The Ethics of Climate Engineering: Solar Radiation Management and Non-Ideal Justice*. New York: Routledge.

Svoboda, T., and Irvine, P.J. (2014). "Ethical and Technical Challenges in Compensating for Harm Due to Solar Radiation Management Geoengineering." *Ethics, Policy & Environment* 17(2): 157–174.

Svoboda, T., Keller, K., Goes, M., and Tuana, N. (2011). "Sulphate Aerosol Geoengineering: The Question of Justice." *Public Affairs Quarterly* 25(3): 157–180.

Szerszynski, B. (2005). *Nature, Technology and the Sacred*. Oxford: Blackwell.

Tadross, M., Suarez, P., Lotsch, A., Hachigonta, S., Mdoka, M., Unganai, L., Lucio, F., Kamdonyo, D., and Muchinda, M. (2009). "Growing-season Rainfall and Scenarios of Future Change in Southeast Africa: Implications for Cultivating Maize." *Climate Research* 40: 147–161.

Tannert, C., Elvers, H., and Jandrig, B. (2007). "The Ethics of Uncertainty." *EMBO Reports* 8(10): 892–896.

Taylor, K.E., Stouffer, R.J., and Meehl, G.A. (2008). "A Summary of the CMIP5 Experiment Design." Available at: http://www.clivar.org/organization/wgcm/references/ Taylor_CMIP5_dec31.pdf

Teich, A.H. (1993). *Technology and the Future, 6th ed.* New York: St. Martin's Press.

Thomas, C.D. et al. 2004. "Extinction Risk from Climate Change." *Nature* 427: 145–148.

Tietsche, S., Notz, D., Jungclaus, J.H., and Marotzke, J. (2011). "Recovery Mechanisms of Arctic Summer Sea Ice." *Geophysical Research Letters* 38(2): L02707.

Tilmes, S., Garcia, R.R., Kinnison, D.E., Gettelman, A., and Rasch, P.J. (2009). "Impact of Geoengineered Aerosols on the Troposphere and Stratosphere." *Journal of Geophysical Research* 114: D12305. doi:10.1029/2008JD011420.

Tilmes, S., Müller, R., and Salawitch, R. (2008). "The Sensitivity of Polar Ozone Depletion to Proposed Geoengineering Schemes." *Science* 320(5880): 1201–1204.

Toffler, A. (1971). *Future Shock*. New York: Random House.

Tollefson, J. (2010). "Geoengineers Get the Fear." *Nature* 464: 656. March 30. Available at: http://www.nature.com/news/2010/100330/full/464656a.html

Tollefson, J. (2012). "Ocean-fertilization Project Off Canada Sparks Furore." *Nature* 490: 458–459. October 23. Available at: http://www.nature.com/news/ocean-fertilization-project-off-canada-sparks-furore-1.11631

Toniazzo, T., Gregory, J.M., and Huybrechts, P. (2004). "Climatic Impact of a Greenland Deglaciation and Its Possible Irreversibility." *Journal of Climate* 17(1): 21–33.

Trenberth, K.E., and Dai, A. (2007). "Effects of Mount Pinatubo Volcanic Eruption on the Hydrological Cycle as an Analog of Geoengineering." *Geophysical Research Letters* 34: L15702. doi: 10.1029/2007GL030524.

Trick, C.G., Bill, B.D., Cochlan, W.P., Wells, M.L., Trainer, V.L., and Pickell, L.D. (2010). "Iron Enrichment Stimulates Toxic Diatom Production in High-nitrate,

Low-chlorophyll Areas." *Proceedings of the National Academy of Sciences* 107(13): 5887–5892.

Tuana, N. (2018). "The Ethical Dimensions of Geoengineering: Solar Radiation Management through Sulphate Particle Injection." In Blackstock, J.J. and Low, S. (eds.). *Geoengineering Our Climate? Ethics, Politics and Governance*. London: Earthscan from Routledge, pp. 71–85 (this volume).

Tuana, N., Sriver, R., Svoboda, T., Olsen, R., Irvine, P., Haqq-Misra, J., and Keller, K. (2012). "Towards Integrated Ethical and Scientific Analysis of Geoengineering: A Research Agenda." *Ethics, Place, and Environment* 15(2): 1–22.

UK Government. (2010). "Government Response to the House of Commons Science and Technology Committee 5th report of session 2009–10: The Regulation of Geoengineering." London: Stationary Office.

United Nations. (1978). "Convention on the Prohibition of Military or Any Other Hostile Use of Environmental Modification Techniques." Available at: http://daccess-ods.un.org/TMP/6130753.html

United Nations. (2008). Millennium Development Goals Indicator. Available at: http://mdgs.un.org/unsd/mdg/SeriesDetail.aspx?srid=749&crid

United Nations Environmental Program (UNEP). (1980). "Provisions for co-operation between States in weather modification: decision 8/7/A of the Governing Council of UNEP of 29 April 1980." Available at: https://digitallibrary.un.org/record/42518

United Nations Framework Convention on Climate Change. (1992). "Preamble Article 1." Available at: http://www.un-documents.net/unfccc.htm

Urban, N.M., and Keller, K. (2010). "Probabilistic Hindcasts and Projections of the Coupled Climate, Carbon Cycle and Atlantic Meridional Overturning Circulation System: A Bayesian Fusion of Century-Scale Observations with a Simple Model." *Tellus Series A-Dynamic Meteorology and Oceanography* 62(5): 737–750.

Urpelainen, J. (2012). "Geoengineering and Global Warming: a Strategic Perspective." *International Environmental Agreements: Politics, Law and Economics* 12(4): 375–389.

US Air Force. (2000). *Military Operations Other Than War*. Air Force Doctrine Document 2–3. Available at: http://www.dtic.mil/doctrine/jel/service_pubs/afd2_3.pdf

US Government Accountability Office. (2009). "Report to the Chairman, Committee on Science and Technology, House of Representatives – Climate Change: A Coordinated Strategy Could Focus Federal Geoengineering Research and Inform Governance Efforts." Report GAO-10-903. Available at: http://www.gao.gov/products/GAO-10-903

US Government Accountability Office. (2010a). *Climate Change: Preliminary Observations on Geoengineering Science, Federal Efforts, and Governance Issues*. Washington, DC: United States Government Accountability Office. Available at: https://www.gao.gov/assets/130/124271.pdf

US Government Accountability Office. (2010b). *Climate Change: A Coordinated Strategy Could Focus Federal Geoengineering Research and Inform Governance Efforts*. Washington, DC: United States Government Accountability Office. Available at: http://www.gao.gov/products/GAO-10-903

US Government Accountability Office. (2011). *Climate Engineering: Technical Status, Future Directions, and Potential Responses*. Washington, DC: United States Government Accountability Office. Available at: http://www.gao.gov/products/GAO-11-71

Van den Hoven, J. (2013). "Value Sensitive Design and Responsible Innovation." In Owen, R., Bessant, J., and Heintz, M. (eds.). *Responsible Innovation: Managing the*

Responsible Emergence of Science and Innovation in Society, First Edition. New York: John Wiley & Sons, pp. 51–74.

Vaughan, N.E., and Lenton, T.M. (2011). "A Review of Climate Geoengineering Proposals." *Climatic Change* 109(3–4): 745–790.

Verbeek, P.-P. (2005). *What Things Do: Philosophical Reflections on Technology, Agency, and Design*. Crease, R. P. (trans.). University Park, PA: Pennsylvania State University Press.

Vervoost, J., and Gupta, A. (2018). "Anticipating Climate Futures in a 1.5C Era: The Link between Foresight and Governance." *Environmental Sustainability* 31: 104–111.

Victor, D.G. (2008). "On the Regulation of Geoengineering." *Oxford Review of Economic Policy* 24(2): 322–336.

Victor, D.G., Morgan, M.G., Apt, J., Steinbruner, J., and Ricke, K.L. (2009). "The Geoengineering Option." *Foreign Affairs* 88(2): 64–76.

Von Schomberg, R. (2013). "A Vision of Responsible Innovation." In Owen, R., Bessant, J., and Heintz, M. (eds.). *Responsible Innovation: Managing the Responsible Emergence of Science and Innovation in Society*, First Edition. New York: John Wiley & Sons, pp. 51–74.

Wack, P. (1985). "Scenarios: Shooting the Rapids." *Harvard Business Review* 63(6): 139–150.

Wadman M. (2004). "Spitzer Sues Drug Giant for Deceiving Doctors." *Nature* 429(6992): 589–589.

Wagner, G., and Weitzman, M. (2012). "Playing God." *Foreign Policy*. October 24. Available at: http://www.foreignpolicy.com/articles/2012/10/22/playing_god

Wagner, G., and Weitzman, M.L. (2015). *Climate Shock: The Economic Consequences of a Hotter Planet*. Princeton, NJ: Princeton University Press.

Wapner, P. (2013). *Living Through the End of Nature: The Future of American Environmentalism*. Cambridge, MA: MIT Press.

Walker, W.E., Rotmans, J., der Sluijs, J.P.V., van Asselt, M.B.A., Janssen, P., and Krauss, M. (2003). "Defining Uncertainty: A Conceptual Basis for Uncertainty Management." *Integrated Assessment* 4(1): 5–17.

Warren, L.F. (1925). *Facts and Plans. Rainmaking—Fogs and Radiant Planes.* Privately Printed. Copy in Wilder Bancroft Papers, Cornell University Library.

Watson, M. (2011). "Testbed Delay." *The Reluctant Geoengineer*. Weblog. September 30. Available at: http://thereluctantgeoengineer.blogspot.co.uk/2011/09/testbed-delay.html

Watson, M. (2012). "Testbed News." *The Reluctant Geoengineer*. Weblog. May 16. Available at: http://thereluctantgeoengineer.blogspot.de/2012/05/testbed-news.html

Watson, M. (2014). Personal communications with Doughty, J.

Weinberg, A.M. (1967). *Reflections on Big Science*. Cambridge, MA: MIT Press.

Weitzman, M.L. (2015). "A Voting Architecture for the Governance of Free-Driver Externalities, with Application to Geoengineering." *Scandinavian Journal of Economics* 117(4): 1049–1068.

Welch, A., Gaines, S., Marjoram, T., and Fonseca, L. (2012). "Climate Engineering: The Way Forward?" *Environmental Development* 2: 57–72.

Welsh, I., and Wynne, B. (2013). "Science, Scientism and Imaginaries of Publics in the UK: Passive Objects, Incipient Threats." *Science as Culture* 22(4): 540–566.

White, L.T. (1967). "The Historical Roots of Our Ecologic Crisis." *Science*, 155(3767): 1203–1207.

Whyte, K.P. (2012). "Now This! Indigenous Sovereignty, Political Obliviousness and Governance Models for SRM Research." *Ethics, Policy & Environment* 15(2): 172–187.

Wiertz, T. (2015). "Visions of Climate Control: Solar Radiation Management in Climate Simulations." *Science, Technology and Human Values.* 41(3): 438–460. doi: 10.1177/0162243915606524.

Wigley, T.M.L. (2006). "A Combined Mitigation/Geoengineering Approach to Climate Stabilization." *Science* 314: 452–454.

Wigley, T. (2011). "Geoengineering the Climate: A Southern Hemisphere Perspective." A Symposium organised by the National Committee for Earth System Science of the Australian Academy of Science.

Wikman-Svahn, P. (2012). "Radiation Protection Issues Related to the Use of Nuclear Power." *WIREs Energy Environ* 1(3): 256–269.

Williamson, P., Wallace, D.W.R., Law, C.S., Boyd, P.W., Collos, Y., Croot, P., Denman, K., Riebesell, U., Takeda, S., and Vivian, C. (2012). "Ocean Fertilization for Geoengineering: A Review of Effectiveness, Environmental Impacts and Emerging Governance." *Process Safety and Environmental Protection* 90(6): 475–488.

Wilsdon, J., and Willis, R. (2004). *See-through Science: Why Public Engagement Needs to Move Upstream.* London: Demos.

Wiltshire, A., and Davies-Barnard, T. (2015). "Planetary limits to BECCS negative emissions." Available at: http://avoid-net-uk.cc.ic.ac.uk/wp-content/uploads/delightful-downloads/2015/07/Planetary-limits-to-BECCS-negative-emissions-AVOID-2_WPD2a_v1.1.pdf

Wisner, B., Blaikie, P., Cannon, T., and Davis, I. (2005). *At Risk: Natural Hazards, People's Vulnerability and Disasters.* London: Routledge.

Wood, D. (2003). "Albert Borgmann on Taming Technology: An Interview." *Christian Century.* August 23: 22–25.

World Intellectual Property Organization (WIPO). (2010a). "Managing IP at CERN." December. Available at: http://www.wipo.int/wipo_magazine/en/2010/06/article_0003.html

World Intellectual Property Organization (WIPO). (2010b). "WIPO and CERN Sign Cooperation Agreement." PR/2010/653. Available at: http://www.wipo.int/pressroom/en/articles/2010/article_0027.html

World People's Conference on Climate Change and the Rights of Mother Earth. (2010). "The Peoples Agreement." Available at: http://pwccc.wordpress.com/support/

Wunsch, C., and Ferrari, R. (2004). "Vertical Mixing, Energy, and the General Circulation of the Oceans." *Annual Review of Fluid Mechanics* 36(1): 281–314.

Zagorin, P. (1998). *Francis Bacon.* Princeton, NJ: Princeton University Press.

Zürn, M., and Schäfer, S. (2013). "The Paradox of Climate Engineering." *Global Policy* 4(3): 266–277.

INDEX